Physiological Function in Special Environments

Charles V. Paganelli and Leon E. Farhi
Editors

Physiological Function in Special Environments

With 84 Illustrations

Springer-Verlag
New York Berlin Heidelberg
London Paris Tokyo

Charles V. Paganelli
Leon E. Farhi
Department of Physiology
State University of New York at Buffalo
Buffalo, NY 14214
USA

The illustration on the front cover was designed by Donald Watkins and the Educational Communications Center, Buffalo, New York, and used as the logo for the Satellite Symposium on Environmental Physiology, Buffalo, October 10–12, 1985.

Library of Congress Cataloging-in-Publication Data

Environmental physiology / editors, Charles V. Paganelli and Leon E. Farhi
    p. cm.
   Bibliography: p.
   Includes index.
   ISBN 0-387-96833-4
   1. Physiology, Comparative—Congresses. 2. Adaptation
(Physiology)—Congresses. I. Paganelli, Charles V.
II. Farhi, Leon E.
QP33.E58 1988
591.5—dc19   88-24892
                                                       CIP

Typeset by TCSystems, Inc., Shippensburg, Pennsylvania
Printed and bound by Arcata Graphics/Halliday, West Hanover, Massachusetts.
Printed in the United States of America.

9 8 7 6 5 4 3 2 1

ISBN 0-387-96833-4 Springer-Verlag New York Berlin Heidelberg
ISBN 3-540-96833-4 Springer-Verlag Berlin Heidelberg New York

Preface

The numerous ways in which man and animals are affected by their physical environment, and the inborn and adaptive responses to change in the "milieu exterieur" have fascinated curious minds since the earliest days of recorded history. Development of the scientific method with its emphasis on evidence obtained through experimentation—perhaps best illustrated in this field by Paul Bert's encyclopedic work—allowed several generations of our predecessors to establish firmly some facts and reject erroneous beliefs, but it was only during the early 1940s that environmental physiology put on its seven-league boots.

In 1941, a young physiologist named Hermann Rahn was recruited by Wallace O. Fenn, then Chairman of the Department of Physiology at the University of Rochester, who was engaged in a study of the effects of altitude on human performance. The years that followed witnessed some of Hermann Rahn's early achievements not only in the area of altitude, but in other aspects of environmental physiology as well. In particular, he participated in the definitive studies of human adaptive mechanisms in arid climates which formed the basis of Edward Adolph's classic "Physiology of Man in the Desert" (Wiley/Interscience, NY 1947). During those golden years, environmental physiology flourished, and important discoveries were reported in a seemingly endless stream from many laboratories. From the perspective of more than 40 years, however, it seems fair to say that many, if not most, of the fundamental aspects of altitude and aviation physiology were described and analyzed by the Rochester triumvirate of Fenn, Rahn, and Otis from work carried out at the bench, in altitude chambers, and in field expeditions to remote places.

After 15 years in Rochester, during which Hermann Rahn rose to the rank of Vice Chairman of the Department, he assumed the leadership of the Department of Physiology at the University of Buffalo in 1956. Within a few years, he had gathered around him a team of younger investigators, many of whom had a demonstrated interest in diverse aspects of environmental physiology. Among the areas that were investigated over the following decade—with the support of such funding agencies as the National Science Foundation, the National Institutes of Health, the U.S. Navy Office of Naval Research, and the U.S. Air Force Office of Scientific Research—were the effects of altitude, diving, temperature, and gravity on human performance. Following the tradition started in

Rochester, research went hand-in-hand with training: scientists from the U.S. and around the world joined the Buffalo group for poriods ranging from several months to three years to participate in its work and "learn by doing."

Against this background, a symposium on environmental physiology was organized in Buffalo in October of 1985 to honor Hermann Rahn, and was the occasion for two other events celebrating his career in environmental physiology. By order of the Council of the University at Buffalo, the environmental physiology facility at our university was officially dedicated to Hermann Rahn, and named The Hermann Rahn Laboratory of Environmental Physiology. The second honor came from the U.S. Air Force in the form of the Meritorius Civilian Service Award, presented to him by Dr. Billy Welch, Chief Scientist, Human Systems Division, Brooks Air Force Base, at the symposium banquet.

The symposium itself took place on October 11 and 12, 1985 as a satellite of the American Physiological Society's Fall Meeting in Niagara Falls, and brought together an international group of 25 participants from the United States, Japan, West Germany, France, Israel, and Switzerland. The chapters in this volume were presented at that symposium; they cover a wide range of topics in environmental physiology, grouped into four general categories: Adaptation to Altitude; Diving and Exposure to Elevated Pressure; Exposure to Altered G-Force; and Comparative Physiology. In addition, Dr. Ewald Weibel, Professor of Anatomy and Director of the Anatomisches Institut, Universität Bern, provided an introductory chapter entitled "Fried Eggs on a Flying Saucer: Exploring the Pathway for Oxygen and Its Environment," originally an imaginative and beautifully illustrated lecture given at the symposium banquet, and intended, in the author's words, as a "surrealistic extension" of certain aspects of Hermann Rahn's many careers in physiology.

The present volume can obviously make no claim to all-inclusiveness. The field of environmental physiology is much too large to have been encompassed in a meeting lasting two days, as witness the many monographs and voluminous literature on the subject. The American Physiological Society's treatment of the topic in its *Handbook of Physiology* series alone occupies two volumes (one issued in 1964 and the second in 1977) and over 1,600 pages. Our selection of topics from a wealth of choices was dictated primarily by the fields of endeavor in which Hermann Rahn has made important, and in many cases, definitive, contributions. (Even so, we could not include temperature regulation for lack of time.)

We were aided in organizing the symposium and in our editorial work on the manuscripts by the Chairmen of the four sessions of the Symposium: Dr. S. Marsh Tenney, Nathan Smith Professor Emeritus of Physiology, Dartmouth Medical School (Altitude), Dr. Tetsuro Yokoyama, Professor of Medicine, Keio University School of Medicine,

Tokyo (Diving and Elevated Pressure); Dr. Arthur H. Smith, Professor Emeritus of Animal Physiology, University of California/Davis (G-Forces); and Dr. Claude Lenfant, Director, National Heart, Lung & Blood Institute, National Institutes of Health (Comparative Physiology). We wish to express to them our most sincere thanks for their efforts, without which the symposium could not have come to fruition. We should also like gratefully to acknowledge the financial support provided by the School of Medicine & Biomedical Sciences, University at Buffalo; by the American Physiological Society; by the Conferences in the Disciplines Program of the Graduate School, University at Buffalo; and by the Department of Physiology, University at Buffalo. Our warm thanks are also due to Mrs. Willie Brownie and Mrs. Julie Maciejewski who provided expert organizing and secretarial help, to Mr. Donald E. Watkins, of the Educational Communications Center, University at Buffalo, for his elegant design of the symposium logo which appears on the cover of this volume, and to the editorial and production staff of Springer-Verlag, New York, for patient assistance in bringing this book manuscript to publication. We particularly wish to express sincere gratitude to our colleagues in the Physiology Department at Buffalo for their help and encouragement in arranging the many details attendant on the symposium, and for their enthusiastic participation in the scientific sessions.

Finally, to Hermann Rahn, for his guidance and inspiration, we dedicate this volume with our admiration and thanks.

Charles Paganelli
Leon Farhi
Buffalo, New York
February 24, 1989

Contents

x   Contents

Contributors

AMOS AR
Department of Zoology, George S. Wise Faculty of Life Sciences, Tel
Aviv University, Tel Aviv, Ramat Aviv 69978, Israel

R.R. BURTON
Crew Technology Division, USAF School of Aerospace Medicine,
Brooks, AFB, Texas 78235-5301, USA

P. CERRETELLI
Department of Physiology, Centre Médical Universitaire, 1, rue
Michel-Servet, 1211 Genève 4, Switzerland

PIERRE DEJOURS
Laboratoire d'Etude des Regulations Physiologiques, associé à
l'Université Louis Pasteur, CNRS, 23 rue Becquerel, 67087 Strasbourg,
France

JEROME A. DEMPSEY
John Rankin Laboratory of Pulmonary Medicine, Department of
Preventive Medicine, University of Wisconsin Medical School,
Madison, Wisconsin 53706, USA

P.E. DI PRAMPERO
Department of Physiology, Centre Médical Universitaire, 1, rue
Michel-Servet, 1211 Genève 4, Switzerland

KONRAD FALKE
Department of Anesthesia, University of Berlin, D 1000 Berlin 19,
West Germany

LEON E. FARHI
Department of Physiology, State University of New York at Buffalo,
Buffalo, New York 14214, USA

J. FASULES
Cardiovascular Pulmonary Research Laboratory B-133, University of
Colorado Health Sciences Center, 4200 E. 9th Avenue, Denver,
Colorado 80262, USA

H.T. HAMMEL
Department of Physiology, Meyers Hall, Indiana University,
Bloomington, Indiana 47405, USA

ROGER D. HILL
Department of Anesthesia, Massachusetts General Hospital, Boston,
Massachusetts 02114, USA

PETER W. HOCHACHKA
Department of Zoology, University of British Columbia, Vancouver,
British Columbia, Canada V6T 2A9

D.C. HOFFMAN
Department of Physics and Astronomy, University of Hawaii,
Honolulu, Hawaii 96822, USA

SUK KI HONG
Department of Physiology, State University of New York at Buffalo,
Buffalo, New York 14214, USA

H. HOWALD
Department of Physiology, Centre Médical Universitaire, 1, rue
Michel-Servet, 1211 Genève 4, Switzerland

PHILIP C. JOHNSON, JR.
NASA/Johnson Space Center, Biomedical Laboratories Branch, Mail
Code SD4, Houston, Texas 77058, USA

SUKHAMAY LAHIRI
Department of Physiology, University of Pennsylvania School of
Medicine, Philadelphia, Pennsylvania 19104, USA

CAROLYN S. LEACH
NASA/Johnson Space Center, Biomedical Laboratories Branch, Mail
Code SD4, Houston, Texas 77058, USA

GRAHAM C. LIGGINS
National Women's Hospital, Postgraduate School of Obstetrics and
Gynecology, Auckland 3, New Zealand

Y. OHTA
Department of Medicine, School of Medicine, Tokai University,
Bohseidai, Isehara, Japan 259-11

CHARLES V. PAGANELLI
Department of Physiology, State University of New York at Buffalo,
Buffalo, New York 14214, USA

JOHANNES PIIPER
Abteilung Physiologie, Max-Planck-Institut für experimentelle Medizin,
Hermann-Rein-Str. 3, 3400 Göttingen, West Germany

JESPER QVIST
Herlev Hospital, Department of Anesthesia, Copenhagen, DK 2730
Herlev, Denmark

HERMANN RAHN
Department of Physiology, State University of New York at Buffalo,
Buffalo, New York 14214, USA

J.T. REEVES
Cardiovascular Pulmonary Research Laboratory B-133, University of
Colorado Health Sciences Center, 4200 E. 9th Avenue, Denver,
Colorado 80262, USA

ROBERT C. SCHNEIDER
Department of Anesthesia, Massachusetts General Hospital, Boston,
Massachusetts 02114, USA

CURTIS A. SMITH
John Rankin Laboratory of Pulmonary Medicine, Department of
Preventive Medicine, University of Wisconsin Medical School,
Madison, Wisconsin 53706, USA

K.R. STENMARK
Cardiovascular Pulmonary Research Laboratory B-133, University of
Colorado Health Sciences Center, 4200 E. 9th Avenue, Denver,
Colorado 80262, USA

A. TUCKER
Department of Physiology, Colorado State University, Fort Collins, Colorado 80523, USA

HUGH D. VAN LIEW
Department of Physiology, State University of New York at Buffalo, Buffalo, New York 14214, USA

EWALD R. WEIBEL
University of Bern, Buhlstr. 26, Postfach 139, Bern 9, Switzerland

D.E. YOUNT
Department of Physics and Astronomy, University of Hawaii, Honolulu, Hawaii 96822, USA

DAVID ZACKS
Department of Zoology, George S. Wise Faculty of Life Sciences, Tel Aviv University, Tel Aviv, Ramat Aviv 69978, Israel

WARREN M. ZAPOL
Department of Anesthesia, Massachusetts General Hospital, Boston, Massachusetts 02114, USA

Chairpersons of the Sessions

Satellite Symposium on Environmental Physiology, Buffalo, October 10–12, 1985

I. Physiology of Adaptation to Altitude
Stephen M. Tenney
Department of Physiology
Dartmouth Medical School
Hanover, New Hampshire 03756, USA

II. Physiology of Diving and Exposure to Elevated Pressure
Tetsuro Yokoyama
Department of Medicine
Keio University School of Medicine
35 Shinanomachi, Shinjuku-ku
Tokyo 160, Japan

III. Physiology of Exposure to Altered G-Force
Arthur H. Smith
Department of Animal Physiology
University of California, Davis
Davis, California 95616, USA

IV. Comparative Physiology
Claude Lenfant
National Heart, Lung and Blood Institute
NIH Bldg 31
Bethesda, Maryland 20892, USA

Fried Eggs on a Flying Saucer: Exploring the Pathway for Oxygen and Its Environment

Banquet speech at the Symposium on Environmental Physiology, dedicated to Hermann Rahn

EWALD R. WEIBEL

"Romance, relevance, and diving" is the title Hermann Rahn chose for his acceptance speech for the Behnke Award. "Fried eggs on a flying saucer"—this title also has to do with relevance and romance as they may relate to Hermann Rahn's vast oeuvre, which began with as relevant topics as the anatomical study of the chick pituitary, but then went on to look at rattlesnakes, water dogs, diving amas, lungfish, bird embryos, and all aspects of the environment from high to low pressure and weightlessness. But I noticed some missing points. And so I thought I should attempt something like a surrealistic extension of this oeuvre, primarily by considering some of the enigmatic beauty that nature has associated with several of its aspects.

In my view Hermann Rahn is a true naturalist who knows no limit to his inquisitiveness. As we all know, he is a bird watcher, an avid observer of everything that flies around—and I would certainly not be surprised if he were one of the few who have ever seen a flying saucer over the horizon. And I would not be surprised either if he were to cherish the romantic idea of boarding one of these flying objects that know no limit for pervading the environment or even diving into the universe, because they must be fascinating vantage points from where to take a new look at our world, a look beyond diving or flying in weightlessness.

For a bird watcher it is natural that he should be concerned with their eggs, and we all know of Hermann Rahn's extended inquiries into how eggs breathe (Rahn et al. 1979). But there is one extension of his work on eggs which is of evident relevance for practical life. This is the fact that small birds have small eggs, and big birds big eggs. It is evident that a fried ostrich egg would be much more spectacular than a fried hummingbird egg. But there is one problem which Hermann Rahn has realized, and this is that big eggs have a thicker shell, more difficult to crack. So he measured the yield point of the shell of eggs of different mass, i.e., the minimum force F required to break the shell (Ar et al. 1979). The significant result is that big eggs offer a great advantage. Since the minimum force F increases with egg mass to the power 0.9, the relative force required to crack an egg—i.e., the force required per mass of egg that falls into the frying pan—this force is twice as large in a hummingbird egg as compared with an ostrich egg.

The first message is clear, particularly if one were to prepare for an expedition on a flying saucer: big eggs are more economical than small ones.

The vantage point of an imaginary expedition on a flying saucer would in fact be significant for Hermann Rahn's approach as a naturalist, which is to be an observer of nature in all its aspects, to look keenly, inquisitively at every detail, and then to put it into perspective. Here began my problems in preparing for this talk: in trying to define his perspective. I thought it might be expressible in three-dimensional space—but no way. In this situation Hermann Rahn would say: no problem, just take the O_2-CO_2 diagram and everything becomes simple, because every point on it has a multidimensional perspective (Rahn and Fenn 1955). This device is indeed rather impressive, and I had wondered how anyone could invent it, until I realized that Hermann Rahn may have come up with the concept when watching birds. Specifically peacocks whose tail fan has an astonishing likeness to some of the curves derived with the O_2-CO_2 diagram (Fig. 1). Indeed we find that the eyes of the tail fan are distributed along well-defined curves, at the intersection of equiangular or logarithmic spirals, Bernoulli's spira mirabilis, the characteristic growth curve of the nautilus.

Beyond this aesthetic quality, the O_2-CO_2 diagram is also quite useful as it presents the pathway for O_2 from inspired air to the cells in a rather compact way. However, I must confess that I also find it rather boring, and perhaps misleadingly simple. It does away with too much of the intricate complexity, of the beauty of the system that makes it possible for the points to be so neatly arranged. What matters for me is what is between the points. This is what I would like to pull into the picture now, by sketching out the design of the respiratory system, following the pathway for O_2 from the lung to the mitochondria (Weibel 1981), and by considering how some of its design properties relate to its functional performance (Fig. 2).

The model on which we operate is closely related to the O_2-CO_2 diagram and to the concept of a PO_2 cascade as a driving force for O_2 flow (Fig. 3). The basic notions are these:

1. The rate at which O_2 must flow through the system is determined by the rate at which it is consumed in the mitochondria, primarily the mitochondria in muscle when we talk about high flow rates close to the limit of aerobic performance.
2. At each level of the system the flow rate is determined by a PO_2 difference as a driving force and by a conductance which is in part determined by design properties.

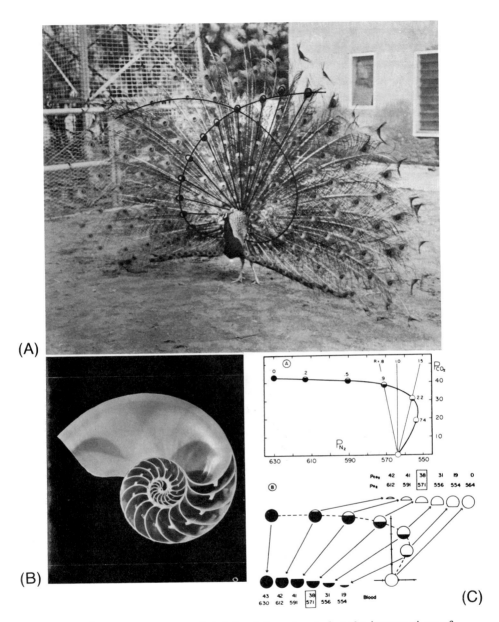

FIGURE 1. The eyes on a peacock tail fan (*a*) are located at the intersections of equiangular spirals, the characteristic growth curve of the nautilus (*b*). Gas tension curves of Rahn (*c*) for comparison (Used by permission of Hermann Rahn).

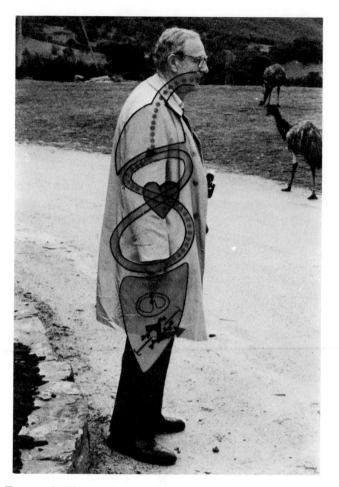

FIGURE 2. The respiratory system and a famous bird watcher.

3. The system is well designed if the conductances are matched to the functional flow requirements under limiting conditions of O_2 consumption. We have called this concept "symmorphosis." (Weibel and Taylor).

I would like to briefly discuss three levels: the lung, the mitochondria, and the capillaries in muscle (Fig. 4). Let me begin with the mitochondria, the sinks into which O_2 disappears in the process of oxidative phosphorylation. When looking at a large variety of animals, including trained and untrained humans, we observed that the total volume of mitochondria in muscle is directly proportional to VO_2 max. This suggests that it is the mitochondria that limit O_2 consumption and by that aerobic metabolism (Hoppeler and Lindstedt 1985). We have looked at this more closely.

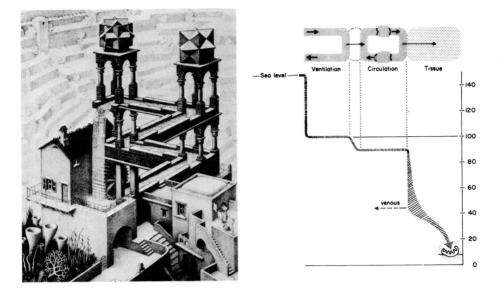

FIGURE 3. Hermann Rahn's PO$_2$ cascade (1962) (used by permission of Hermann Rahn) and M. C. Escher's miraculous Waterfall of 1961. © 1988 M. C. Escher Heirs/Cordon Art-Barn-Holland)

Aerobic metabolism, or rather oxidative phosphorylation, takes place in the inner membrane of the mitochondria (Fig. 4d), where a complex of rather densely arranged respiratory or energy-transducing enzymes entertains a flux of electrons within the membrane, coupled to a flux of protons across the membrane, which, in the end, makes ATP and uses O$_2$ at the cytochrome oxidase (Fig. 4e). By measuring enzyme densities in these mitochondria (Schwerzmann 1986), we have recently estimated that this system operates in vivo close to its maximal reaction rate, namely, at about 24 oxidations per second per cytochrome oxidase when muscles work at their maximal aerobic capacity. This is strong evidence that O$_2$ consumption is indeed limited by the amount of mitochondria that can perform oxidative phosphorylation. So at the last step of the respiratory system the flow of O$_2$ is determined by the size of the sink, or by the conductance at this level.

What about the next level, the delivery of O$_2$ to the cells which must be related to the capillaries (Fig. 4c), as we suspect since A. Krogh's pioneering work. Are capillaries matched to the O$_2$ needs of the cells? Apparently, because we find over a wide range of muscles (from cow locomotor muscles to the myocardium of the shrew) a constant volume ratio of mitochondria to capillary blood, namely 3 ml mitochondria per 1 ml blood. Let us look at what this means. By combining physiological with morphometrical measurements, Hans Hoppeler has estimated that

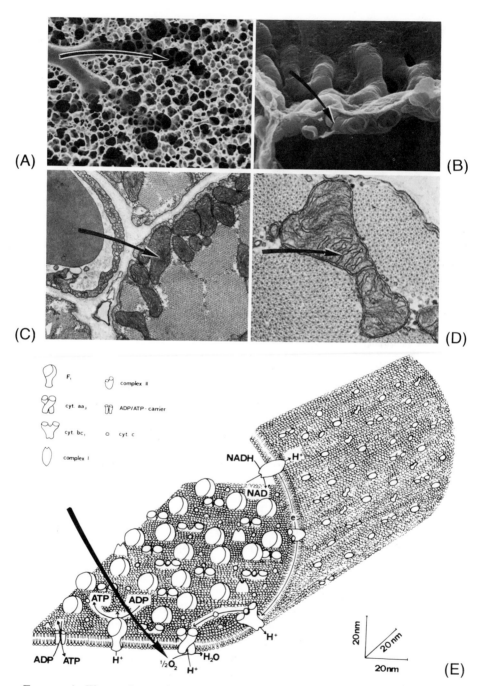

FIGURE 4. The pathway for oxygen (arrows) from the lung (*a, b*), through capillaries (*c*), to muscle mitochondria (*d*), where O_2 is consumed on the molecular complexes of the inner membrane (*e*).

these 3 ml of mitochondria consume 11 μmol $O_2 \cdot sec^{-1}$ when they operate at maximum (Hoppeler and Lindstedt 1985). On the other hand, 1 ml of arterial blood contains 8–9 μmol O_2, of which 70%, or 6 μmol, can be discharged to the cells while the blood flows through the capillaries. To deliver 11 μmol·sec,$^{-1}$ the blood must flow through with a transit time of about 0.5 sec, which is what one finds by direct observation. So it appears that at the level of muscle cells the design properties are well matched to the flow requirements of the system.

Let us finally look at the entrance to the system, the lung, where the conductance for O_2 flow from air to blood is evidently the diffusing capacity D_L. The design characteristics that determine D_L (Figs. 4a and b) are an enormous surface for air-blood contact—of the acreage of a tennis court in man—and a very thin tissue barrier, from which we estimate, for a young adult, D_L to be about 200 ml $O_2 min^{-1} \cdot Torr^{-1}$, about twice as much as one estimates from physiological data on exercise. Is D_L too large? In recent studies with C. R. Taylor we found that D_L is a limiting factor for 0_2 uptake in the running dog or horse, whereas we humans or goats seem to have excess diffusing capacity of the lung, more than we need to ensure maximal O_2 uptake. This statement relates to "normal man," whereas in highly trained athletes, whose O_2 consumption is up to twice as high as with us, D_L may be just right. And furthermore, some reserve D_L at sea level will allow us to climb mountains without severely limiting our work capacity.

With respect to estimating D_L from morphometrical data, there has been some debate, but recent evidence suggests that it is not far from reality, although morphometry consistently yields higher values than physiology (Weibel et al. 1983). This brings us back to eggs, because a few years ago Douglas Wangensteen and I estimated, at the instigation of Hermann Rahn, the diffusing capacity of the chick chorioallantois using a slight modification of the morphometrical model (Wangensteen and Weibel 1982), and we obtained a value very similar to that estimated by Piiper, Tazawa, Ar, and Hermann Rahn on the basis of physiological data (Piiper et al. 1980). This is but one example of the great potential of a comparative approach when we try to explore the foundations of a functional system of great complexity, and Hermann Rahn is one of the protagonists of this approach. The many variables related to this system require that we look for as many solutions, and this we can achieve in the most natural way if we exploit the adaptive variations of evolution. Even the wealth offered by the adaptability of mammals to different environments, to different strategies of life, to different sizes, is enormous.

However, one of the most intriguing adaptive variations relates to the fundamental change in environment, the transition from breathing water to breathing air. For an artist such as M. C. Escher the transformation from fish to bird is simple but not less attractive (Fig. 5). Evolution had to

FIGURE 5. *Sky and Water I* (1938) by M. C. Escher (© 1988 M. C. Escher Heirs/Cordon Art-Barn-Holland).

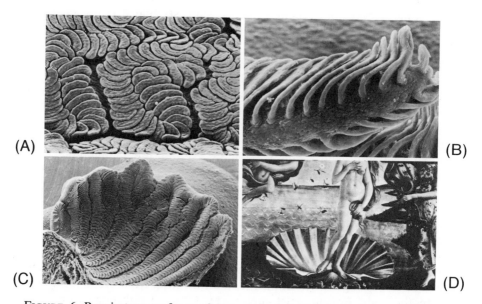

FIGURE 6. Respiratory surfaces of *Heteropneustes fossilis* in the air sac (*a*), on gill filament (*b*), and on the gill fan (*c*) whose design compares well with Botticelli's scallop (*d*).

overcome more difficult problems, of which the strategies for regulating acid-base balance is but one example.

One of the fascinating fields of research is to look at how fish adapt the design of their respiratory apparatus when they need to get O_2 from air as well as from water, as it occurs in lungfish from South America, Africa, or India, when periods of drought force the fish to spend some time onshore (Rahn et al. 1971). It is interesting that at least 3–4 totally different strategies are used to generate an air-breathing surface from available structural resources. One of the most attractive fish to observe is *Heteropneustes fossilis,* the Indian air-breathing catfish, which we have had the pleasure to study with Jyoti Datta-Munshi. This fish retains its gills for water breathing but then forms two long air sacs derived from the gill chambers which extend dorsally on either side of the spine. It is interesting that the air-breathing surface structures are derived from the gill structure by transformation of gill filaments, and that the entrance to the air sac is provided with a valve, the gill fan, which is derived from the last gill arch.

If we first look at the respiratory structures in the air sac (Fig. 6a), we find that they consist of arrays of lamellae as they constitute the surface of gill filaments (Fig. 6b), but that these lamellae are shortened and flattened down so as to avoid an overlap of neighboring gas-exchange units, a rather ingenious adaptation of design to altered functional requirements. But I should not hide from you that I am particularly fascinated by the beautiful design achieved in making the gill fan (Fig. 6c), the valve that allows the air sac to be closed off toward the gill chamber. It is evident that nature has invented a few fundamental forms of mathematical beauty and that she surprises a keen observer by hiding them away in places where one would not expect them.

FIGURE 7. Observers of the Great Trip (Leon Farhi and Michael Hlastala).

So I have nearly closed the circle. Early on I related the O_2-CO_2 diagram to the peacock tail fan and to the equinangular spiral of the nautilus (Fig. 1). The form of the scallop obeys the same basic law of design: that of symmetric, well-proportioned growth, a fundamental, though often complex, law which I like to believe governs the design of well-functioning living creatures, even though we may not easily discover it with our modest means.

But for you, Hermann Rahn, the message is that there is still much enigmatic beauty to search out and to understand, beauty hidden away not only in the depth of lungfish or birds, but anywhere in our world. I suggest that you try to take a new, fresh look at this world, and why not plan for an expedition on a flying saucer, taking some big eggs along to fry? In fact, I am certain that Leon Farhi and others of your friends will be there to see you fly off (Fig. 7).

REFERENCES

Ar A, Rahn H, Paganelli CV (1979). The avian egg: Mass and strength. *Condor* 81: 331–337

Hoppeler H, Lindstedt SL (1985). Malleability of skeletal muscle in overcoming limitations: Structural elements. *J Exp Biol* 115: 355–364

Piiper J, Tazawa H, Ar A, Rahn H (1980). Analysis of chorioallantoic gas exchange in the chick embryo. *Respir Physiol* 39: 273–284

Rahn H, Ar A, Paganelli CV (1979). How bird eggs breathe. Sci Am 240: 46–55

Rahn H, Fenn WO (1955). *A Graphical Analysis of the Respiratory Gas Exchange* (revision of WADC Technical Report 53-255, 1953). The American Physiological Society, Washington, DC

Rahn H, Rahn KB, Howell BJ, Gans C, Tenney SM (1971). Air breathing of the garfish (*Lepisosteus osseus*). *Respir Physiol* 11: 285–307

Schwerzmann K, Cruz-Orive LM, Eggmann R, Sänger A, Weibel ER (1986). Molecular architecture of the inner membrane of mitochondria from rat liver: A combined biochemical and stereological study. *J Cell Biol* 102: 97–103

Wangensteen D, Weibel ER (1982). Morphometric evaluation of chorioallantoic oxygen transport in the chick embryo. *Respir Physiol* 47: 1–20

Weibel ER, (1984). *The pathway for oxygen*. Harvard University Press, Cambridge, Massachusetts

Weibel ER, Taylor CR (1981). Design of the mammalian respiratory system. *Respir Physiol* 44: 1–164

Weibel, ER, Taylor CR, O'Neil JJ, Leith DE, Gehr P, Hoppeler H, Langman V, Baudinette RV (1983). Maximal oxygen consumption and pulmonary diffusing capacity: A direct comparison of physiologic and morphometric measurements in canids. *Respir Physiol* 54: 173–188

Part I Physiology of Adaptation to Altitude

Structure and Function of Carotid Bodies at High Altitude

SUKHAMAY LAHIRI

The mechanism of respiratory drive during acclimatization to environmental hypoxia is a much debated question. Rahn and Otis (1949) formulated and summarized the question in an alveolar P_{O_2} - P_{CO_2} diagram, which is partially reproduced in Fig. 1.1. This figure shows alveolar ventilatory responses in humans to acute (broken lines) and chronic (solid line) exposure to altitudes. The altitude diagonals for sea level, 10,000 ft and 18,000 ft (RQ = 0.85), are also shown. The solid line connecting A and B represents the alveolar values of the acclimatized newcomers to high altitude. Point A is the sea-level value, and the dotted line originating at this point is the response to acute hypoxia. Point B is the alveolar value for humans acclimatized at approximately 10,000 ft, and the dotted line passing through B again indicates the ventilatory responses to acute changes in altitude. If a man were transported to 10,000 ft from sea level, his alveolar pathway would start at A and proceed along the dotted line to the intersection with the 10,000-ft diagonal. With time it would travel down the 10,000-ft altitude diagonal to its intersection with the acclimatized curve (point B). If the man returned to sea level, his alveolar pathway would travel from B on the dotted line to the right until it intersects the sea-level diagonal, and from there proceed to point A along the diagonal in several days. This description of the effects of hypoxia in humans has amply been confirmed in all its essential elements (Lahiri 1968). However, the mechanism of the effects is still unclear. It is well-known now that the initial stimulation of ventilation during acute hypoxia is mediated by the peripheral chemoreceptors (see Lahiri et al. 1983). Without this initial chemoreflex response, ventilatory acclimatization does not seem to occur (Forster et al. 1981; Lahiri et al. 1981). However, what effects chronic hypoxia exert on carotid body structure and function to bring this function about are not known.

As a biological rule, the cellular processes by which the chemoreception occurs are expected to change as a result of the persistent hypoxic stimulus. An expression of this change can be found in the structure and in the metabolism of the putative neurotransmitter. A larger size of the organ has been documented in several species which were exposed to hypoxia for various periods—a few weeks to several years

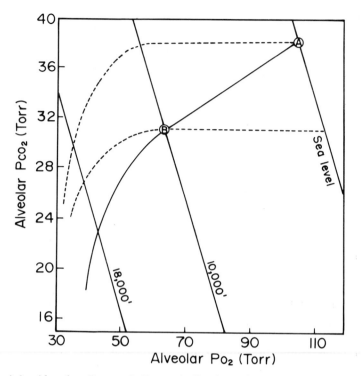

FIGURE 1.1. Alveolar Po₂ and Pco₂, indicating alveolar ventilation, in the acclimatized human (solid line joining A and B), and the changes which occur upon acute exposure to lower oxygen tension for humans at sea level (dotted line, originating at A) and to lower or higher oxygen tension at 10,000 ft (dotted line, originating at B). Altitude diagonals for sea level, 10,000 ft, and 18,000 ft at RQ of 0.85 are shown. (Reproduced from Rahn and Otis, 1949, with permission.)

(Heath and Williams 1977). Ultrastructural studies showed that particularly the glomus cells undergo hypertrophy (McGregor et al. 1984; Pequignot et al. 1984). Concomitantly with the growth of the carotid bodies in the rat during 2 to 4 weeks of chronic hypoxia (inspired Po₂ of 70 Torr), the levels of catecholamines, dopamine, and norepinephrine and their turnover rates are strikingly increased (Hanbauer et al. 1981; Olson et al. 1983). These catecholamines may be involved in the excitation of the chemoreceptor afferents (Fidone et al. 1982), although exogenously administered dopamine is known to be inhibitory (see Eyzaguirre et al. 1983). Accordingly, a reasonable working hypothesis is that these structural and metabolic changes due to chronic hypoxia of several days would affect the chemosensory responses to hypoxia.

Astrand (1954) previously wanted to answer the question whether carotid chemoreceptor activity is decreased after the initial response, and is reset to a lower value during the early phase of ventilatory acclimatization to chronic hypoxia, as suggested by Bjurstedt (1946). Astrand (1954) exposed cats to 4000 m for up to 64 hr and made qualitative measurements of carotid body chemosensory activity from the cut sinus nerve, and showed that the chemoreceptors were still active and responsive to changes in arterial PO_2. This study did not address and answer the question of whether the chemosensory responses changed with the changes in the structure and in catecholamine levels. Astrand's study, however, was not designed to quantitate the chemoreceptor responses. Later, Hornbein and Severinghaus (1969) studied carotid body chemosensory responses of cats native to high altitudes (3800 m and 4300 m). They attempted quantitation by measuring the activity of the whole carotid sinus nerve. Since this method measures electrical activities of an unknown number of fibers which make electrical contact with the recording electrode, a direct comparison between the control and experimental groups could not be made. For comparison they normalized the activity as a function of maximal discharge of the respective carotid sinus nerves (CSN) due to transient asphyxia. They found that the shapes of the response curves in high-altitude and sea-level cats were similar. As discussed by the authors, this method only describes the pattern of response. Thus the absolute quantity of discharge rates of the separate nerve fibers was not known. This study, however, noted that the absolute magnitude of the maximal asphyxic response of the whole nerve in the high-altitude cat, measured in microvolts, was three times greater than that in the sea-level cat. This difference in microvolts, as pointed out by the authors, could have been due to a greater activity of the individual nerve fibers or to nonphysiological factors, such as volume conduction between the nerve and the electrode. If physiological, the responses of the individual chemoreceptor afferents to hypoxia would be three times greater in the high-altitude cat. Thus, the study did not clearly answer their question regarding the cat native to high altitudes. The chemoreceptor responses in the cat native to high altitude need not be the same, however, as those acclimatized to hypoxia for a few weeks (Lahiri et al. 1983). Consequently, their question was different from the one which inquires into the discharge rate of individual carotid chemoreceptor afferents in the cat recently acclimatized to chronic hypoxia.

A possible role of CSN parasympathetic efferent in the control of carotid chemoreceptor activity at high altitude is also of interest because the inhibitory effect due to the efferent activity may be greater in chronic hypoxia (Lahiri et al. 1984).

We studied and compared the responses of individual carotid chemosensory fibers in the sea-level cats; one group was exposed to inspired PO_2 of 70–71 Torr for 28 days, and the other (control) group breathed room air at sea level. We studied (a) hypoxic sensitivity of single chemoreceptor afferents in other-wise intact CSN, (b) the effects of transecting the CSN and hence eliminating the efferent inhibition on the chemosensory activity, and (c) the maximal sustained peak activities due to severe hypoxia. We also compared ultrastructure of the sensory nerve endings on glomus cells in the carotid bodies in two groups of cats.

Experimental Procedure

Control and chronically hypoxic cats were anesthetized initially with α-choralose (60 mg·kg-1, intraperitoneal). During anesthesia and surgery the inspired Po_2 of the cats was maintained at the level at which they were exposed before anesthesia. After tracheostomy and intubation, the upper end of the trachea and esophagus were removed to expose the carotid sinus nerve medially. The femoral arteries were catheterized; one was used for sampling arterial blood, and the other was connected to a Statham transducer to monitor arterial blood pressure.

A slip of the left CSN was cut, leaving the main trunk intact, and a single- or a few-fiber preparation was made for measuring the chemoreceptor activity as described previously (Lahiri and DeLaney 1975). The discharge of the individual fibers was selected by a window discriminator and counted by a rate meter. Tracheal O_2 and CO_2 levels were monitored using calibrated Beckman OM-11 and LB-2 analyzers, respectively. The end-tidal Po_2 and Pco_2 were altered and maintained at predetermined levels by adjusting the inspired gases. The levels of ventilation and inspired Pco_2 were adjusted to make the arterial pH values similar in the two groups of cats. Tracheal O_2 and CO_2 levels, arterial blood pressure, and carotid chemoreceptor discharge rate were all continuously recorded. For a steady-state response, constant levels of end-tidal gases were maintained for 5 min. Arterial blood samples (0.5 ml) were taken anaerobically at a predetermined time, and the Po_2, Pco_2, and pH were measured at 38°C.

The experimental protocols were as follows. (a) Carotid chemoreceptor activities at three steady-state levels of Pa_{O_2} at a constant arterial pH were measured in the chronically hypoxic and normoxic control cats. (b) The maximal discharge rates of the afferents were measured by artificially ventilating the paralyzed cats with nitrogen. (c) Effects of transecting the CSN on the chemosensory discharge rate were measured during hypoxia. (d) Carotid bodies from hypoxic and control normoxic cats which were not used for the measurement of carotid chemoreceptor activites were fixed with buffered glutaraldehyde for the ultrastructural studies. We selected those fields which contained nerve endings on glomus cells for electron micrographs. The purpose was to make a qualitative assessment of the effects of chronic hypoxia on the sensory nerve endings on glomus cells.

Results

CHEMOSENSORY RESPONSES TO HYPOXIA

The arterial pH values among the two groups were comparable (7.360), although the Pa_{O_2} values were lower in the chronically hypoxic group

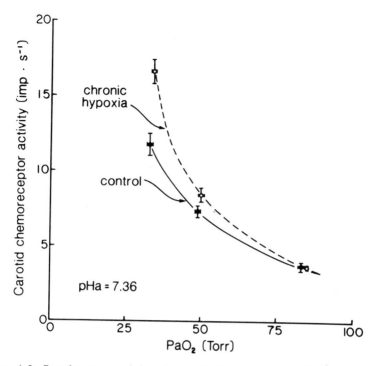

FIGURE 1.2. Steady-state activity of carotid chemoreceptor afferents in normal sea-level cats exposed to air (solid symbol) and to 10% O_2 (open symbol). (Mean ± 1 SD for 16 single afferents in each group.)

(21 Torr vs. 30 Torr). This means that hyperventilation during chronic hypoxia lowered Pa_{CO_2}, causing alkalosis which was mostly compensated by the loss of base excess.

The average steady-state responses of single-carotid chemoreceptors are presented in Fig. 1.2. At high PaO_2, the activities were not different. The responses significantly increased ($P<0.05$), with hypoxia in both experimental and control groups. Comparison of the hypoxic effects between the two groups by the analysis of variance showed a significantly greater increase ($P<0.05$) in the chronically hypoxic group.

MAXIMAL RESPONSE OF THE CHEMORECEPTORS

Upon ventilating the cat with nitrogen, carotid chemoreceptor activity increased promptly to a maximal activity which was sustained for several seconds before showing a decline. An experimental record showing the maximal peak response is presented in Fig. 1.3. A decline in the activity after a plateau ensured that the maximal activity was reached. For comparison, we expressed the average activity per second. The same

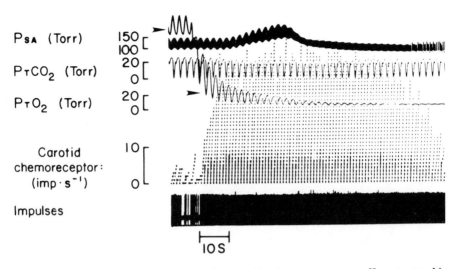

PsA (Torr)

PTCO$_2$ (Torr)

PTO$_2$ (Torr)

Carotid
chemoreceptor:
(imp · s^{-1})

Impulses

10 S

FIGURE 1.3. Maximal response of carotid chemoreceptor afferent to N$_2$ breathing. Tracings from top are arterial blood pressure (PsA), tracheal PCO$_2$ (PTCO$_2$), tracheal PO$_2$ (PTO$_2$, identified by arrows); carotid chemoreceptor activity measured in impulses per second and impulses.

purpose could be achieved by the pulse interval analysis. The average maximal activities (mean ± SD) for the control group ($n = 9$) were $23 ± 6$ imp.sec-1 and for the chronically hypoxic group ($n = 6$), $27 ± 5$ imp.sec-1. The greater mean value in the latter was not significantly ($P>0.05$) different from the control.

MAGNITUDE OF CSN EFFERENT INHIBITION

The effects of transection of CSN of the chemosensory discharge rate during hypoxia are shown in Fig. 1.4. In the control group the CSN transection at Pa$_{O_2}$ of 39 Torr increased the mean activity ($n = 6$) from 11.6 to 12.1 imp.sec-1, whereas in the chronically hypoxic group ($n = 8$) at Pa$_{O_2}$ of 43 Torr, the increment was from 13.6 to 18.0 imp.sec-1. The latter increase was significant ($P<0.01$).

Structure of Carotid Body

Examples of carotid body morphology of normoxic and chronically hypoxic cats are shown, respectively, in Fig. 1.5a and b. The purpose of these micrographs is primarily to draw attention to the structural components of the sensory endings on the glomus cells. The glomus cells in both groups appear similar with respect to mitochondria and dense core

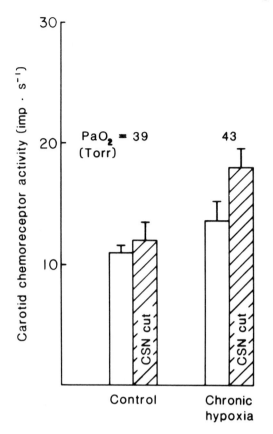

FIGURE 1.4. Effect of carotid sinus nerve transection on the carotid chemorecep-tor activity in the control and chronically hypoxic cats (mean ± 1 SD)

vesicles. There were no gross differences between the two groups. It is of interest, however, that the nerve ending (N) in the chronically hypoxic cat (Fig. 1.5*b*) shows more numerous clear core vesicles which seemed to be in a state of active movement toward the junction with the glomus cell (arrow).

Discussion

The results of the present study show that the responses to hypoxia of the chemoreceptor afferents with intact CSN were greater in the chronically hypoxic than in the control cats without a significant increase in the average maximal discharge rate. This means that the sensitivity of the chemoreceptors to hypoxia increased during chronic stimulation and that the activity provided a greater respiratory drive. Thus, this study

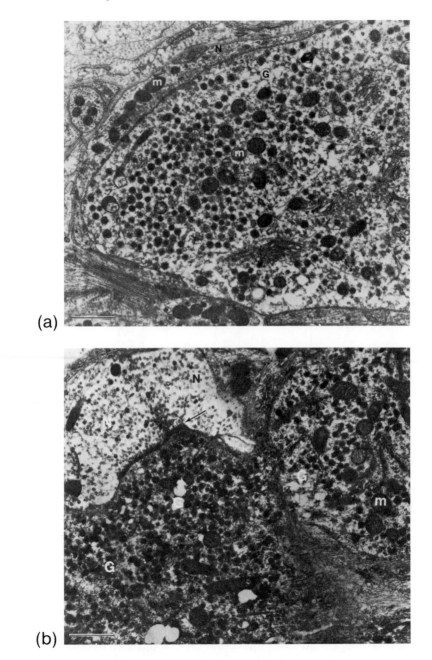

(a)

(b)

FIGURE 1.5. Ultrastructure of cat carotid bodies showing nerve endings in normoxic (*upper panel, a*) and chronically hypoxic (*lower panel, b*) cats. N = nerve ending with numerous clear core vesicles; G = glomus cell; m = mitochondria. Arrow in the lower panel indicates vesicles in the nerve ending swarming toward the junction with a part of a glomus cell. Bar = 1 μm.

provided a more complete answer to the question which was posed by the study of Rahn and Otis (1949) and by many investigators since then (Busch et al. 1985; Forster et al. 1981; Hornbein and Severinghaus 1969; Lahiri et al. 1981). The carotid chemoreceptor sensitivity to hypoxia did not decrease with time. On the basis of this observation it is reasonable to suggest that the peripheral chemoreceptors can contribute to ventilatory acclimatization. However, a direct contribution cannot account for all the ventilation increases during chronic hypoxia because the sensitivity increase was modest and the actual chemosensory traffic to the central nervous system may not have increased. What was gained by the chemoreceptor sensitivity increase was mostly lost due to a rise in Pa_{O_2} through ventilation increases. However, this drive from the peripheral chemoreceptors is critical for sustained high ventilation at altitudes, as we know that chemodenervation leads to depression of ventilation and apnea in anesthetized high-altitude cats (Lahiri 1977). In summary, then, peripheral chemoreceptors are essential to initiate and to sustain ventilatory acclimatization to hypoxia without a substantial increase in the actual chemosensory input to the central nervous system. The increase in ventilation during chronic hypoxia must then be due to a greater central respiratory response to the same chemosensory input. This increased central sensitivity to the peripheral chemoreceptor activity may also be a time-dependent function of the excitatory input to the central nervous system.

It is noteworthy that transection of the CSN increased the discharge rate of the chemoreceptor afferents more in the chronically hypoxic group than in the control. Accordingly, the responses of the chemoreceptors to hypoxia would be even greater after cutting the CSN in the chronically hypoxic cats. These increases, however, were modest, and would not account for a threefold increase compared with the control, as construed by the observations of Hornbein and Severinghaus (1969). They made their observations on the cut CSN.

This present work, however, is not free from sampling problem. The responses of single chemoreceptor afferents in a cat vary considerably (Lahiri and DeLaney 1975). Clearly, all the chemosensory afferents could not be studied in each cat by monitoring the individual fibers one by one and comparing the average results. The fibers which were accepted for the study were selected by trials from many fibers in each cat. The fibers giving weak and unclear signals were discarded, and the procedure was continued until the fibers giving strong and stable signals were found. By the selection process we presumably studied similar types of fibers in the two groups of cats without any particular bias. It is possible but unlikely that the types of fibers which were not studied would have given different results. A study of a larger number of fibers would have increased the confidence, but that would not have presumably altered the validity of the comparison. Alternatively, it would have been ideal to study the same chemoreceptor afferents during the whole period over a few days of chronic hypoxia in order to determine whether the same chemoreceptor changed its response to the same arterial hypoxia with time. Given these limitations and reservations, we may consider that the observed results approximate the actual physiological function.

Although the results define the role of arterial chemoreceptors in the ventilatory acclimatization, the mechanisms leading to a greater excitation of the chemoreceptors by the same arterial stimulus are not revealed by the present experimental results. However, the level of chemoreceptor activity being a balance between the effects of the excitatory and inhibitory factors, one may speculate that the release of an excitatory transmitter was perhaps greater rather than a decrease in the release of an inhibitory factor. This reasoning is based on the observation that the level of dopamine which is inhibitory (see Eyzaguirre et al. 1983) is increased by several folds in the rat carotid body due to chronic hypoxia (Hanbauer et al. 1981; Olson et al. 1983). Its secretion presumably increased also, accounting for the increased efferent inhibition. It is also possible that the efferents to the carotid body are dopaminergic (S. Fidone, personal communication). It is not known, however, whether the dopaminergic activity of these fibers is also increased during chronic hypoxia and what contribution it may make to the observed efferent inhibition (Lahiri et al. 1984). An additional explanation for the augmented efferent inhibition is that the number of the "post-synaptic" receptors for the putative neurotransmitter is increased during chronic hypoxia. This change could be facilitated by an increased surface area of the nerve endings on the glomus cells.

The ultrastructural study showed that the nerve endings on the glomus cells in the two groups were not qualitatively different. If the nerve-ending mitochondria were specifically affected during the acute stage of hypoxia as reported by McDonald (1981) and Hansen (1981), these structural changes did not persist during the chronic state of hypoxia, although a subtle residual effect could have been there. On the other hand, the vesicles in the nerve ending appeared to be more numerous and "active" in the chronically hypoxic group. This may indicate a complete recovery from the depleted stage of acute hypoxia and an adaptive response which may be linked to the observed enhanced chemosensitivity as well as efferent inhibition. The function of these vesicles is not known.

In the context of this paper it is relevant to point out that adult high-altitude natives have been found to show blunted ventilatory response to hypoxia (see Lahiri 1984). The mechanism of this phenomenon is not clear. Tenney and Ou (1977) developed and studied a chronically hypoxic cat model. Their results indicated that cerebral cortex critically contributed to the expression of the phenomenon. We attempted to develop such a cat model to study the carotid chemoreceptors. We found that the carotid chemoreceptor responses to hypoxia were not attenuated even though some of these cats showed blunted ventilatory response to hypoxia (unpublished observations, M. Pokorski, P. Barnard, S. Lahiri). This observation indirectly supports the view expressed by Tenney and Ou (1977).

Clearly, carotid body manifests structural, biochemical, and functional responses to chronic hypoxia. All these responses affirm that carotid body is truly an oxygen-sensitive organ. Whether single or multiple oxygen-sensitive systems operate to bring about the changes remains a challenging unknown. It is hard not to think that the responses initiated by acute hypoxia are not integrated at the cellular level in the chronic state. This cellular response perhaps makes it more suitable for carotid body to play its designated role at high altitude, generating appropriate physiological reflexes for survival.

Acknowledgment. This paper owes much to my associates for their participation at various stages of the work. Supported in part by the NIH Grants HL-19737 and NS 21068.

REFERENCES

Åstrand PO (1954). A study of chemoreceptor activity in animals exposed to prolonged hypoxia. *Acta Physiol Scand* 30: 335–342

Bjurstedt AGH (1946). Interaction of centrogenic and chemoreflex control of breathing during oxygen deficiency at rest. *Acta Physiol Scand (Suppl* 38) 12: 1–88

Busch MA, Bisgard GE, Forster HV (1985). Ventilatory acclimatization to hypoxia is not dependent on arterial hypoxemia. *J Appl Physiol* 58: 1874–1880

Eyzaguirre C, Fitzgerald RS, Lahiri S, Zapata P (1983). Arterial chemoreceptors. In: Shepherd JT, Abbound FM (eds) *The Cardiovascular System (Handbook of Physiology)*. American Physiological Society, Bethesda, MD, sec 3, vol III, pp 557–621

Fidone S, Gonzalez C, Yoshizaki K (1982). Effects of low oxygen on the release of dopamine from the rabbit carotid body *in vitro*. *J Physiol* 333: 93–110

Forster HV, Bisgard GE, Klein JP (1981). Effect of peripheral chemoreceptor denervation on acclimatization of goats during hypoxia. *J. Appl Physiol* 50: 392–398

Hanbauer I, Karoum F, Hellstrom S, Lahiri S (1981). Effects of hypoxia lasting up to one month on the catecholamine content in rat carotid body. *Neuroscience* 6: 81–86

Hansen JT (1981). Chemoreceptor nerve and type A glomus cell activity following hypoxia, hypercapnia or anoxia: A morphological study in the rat carotid body. *J Ultrastruct Res* 77: 189–198

Heath D, Williams DR (1977). *Man at High Altitude*. Churchill Livingstone, Edinburgh

Hornbein TF, Severinghaus JW (1969). Carotid chemoreceptor response to hypoxia and acidosis in cats living at high altitude. *J. Appl Physiol* 27: 837–839

Lahiri S (1968). Alveolar gas pressures in man with life-time hypoxia. *Respir Physiol* 4: 376–386

Lahiri S (1977). Ventilatory response to hypoxia in intact cats living at 3,850 m. *J Appl Physiol* 43: 114–120

Lahiri S (1984). Respiratory control in Andean and Himalayan high altitude natives. In: West JB, Lahiri S (eds) *High Altitude and Man*. American Physiological Society, Bethesda, MD, pp 147–162

Lahiri S, Barnard P, Zhang R (1983). Initiation and control of ventilatory adaptation to chronic hypoxia of altitude. In: Pallot DJ (ed) *Control of Respiration*. Croom Helm, London, pp 298–325

Lahiri S, DeLaney RG (1975). Stimulus interaction in the responses of carotid body chemoreceptor single afferent fibers. *Respir Physiol* 24: 249–266

Lahiri S, Edelman N, Cherniack NS, Fishman AP (1981). Role of carotid chemoreflex in respiratory acclimatization to hypoxemia in goat and sheep. *Respir Physiol* 46: 367–382

Lahiri S, Smatresk NJ, Pokorski M, Barnard P, Mokashi A, McGregor, KH (1984). Dopaminergic efferent inhibition of carotid body chemoreceptors in chronically hypoxic cats. *Am J Physiol* 247 (*Regulatory Integrative Comp Physiol* 17): R24–R28

McDonald DM (1981). Peripheral chemoreceptors. In: Hornbein TF (ed) *Regulation of Breathing*. Marcel Dekker, New York, pp 105–319

McGregor KH, Gil J, Lahiri, S (1984). A morphometric study of the carotid body in chronically hypoxic cats. *J Appl Physiol* 57: 1430–1438

Olson EB Jr, Vidruk EH, McCrimmon DR, Dempsey JA (1983). Monoamine neurotransmitter metabolism during acclimatization to hypoxia in rats. *Respir Physiol* 54: 79–96

Pequignot JM, Hellstrom S, Johansson C (1984). Intact and sympathectomized carotid bodies of long-term hypoxic rats: A morphometric ultrastructural study. *J Neurocytol* 13: 481–493

Rahn H, Otis AB (1949). Man's respiratory response during and after acclimatization to high altitude. *Am J Physiol* 157: 445–462

Tenney SM, Ou LC (1977). Hypoxic ventilatory response of cats at high altitude: An interpretation of blunting. *Respir Physiol* 30: 185–199

2

Where Does [H$^+$] Fit in the Scheme of Ventilatory Acclimatization to Hypoxia?

JEROME A. DEMPSEY AND CURTIS A. SMITH

The Problem

We will address the mediation of short-term ventilatory acclimatization to hypoxia in the sea-level native sojourning at high altitude. This process has already been described in great detail by Dr. Rahn over 35 years ago (Rahn and Otis, 1949), and we reiterate only one example in humans at a single altitude in Fig. 2.1. The characteristics of this process are identical during wakefulness, all intensities of exercise, and all sleep stages; most nonhuman mammals show a very similar time course of change—although usually the magnitude of ventilatory response is completed much more quickly in rats, ponies, goats, and dogs than it is in humans (Berssenbrugge et al. 1984; Dempsey and Forster 1982). In addition, ventilatory acclimatization occurs over a wide range of altitudes from 3000 m (where the acute response may be immeasurable and the chronic response consists of only moderate hypocapnia) to greater than 6000 m (where the acute response reduces Pa_{CO_2} to less than 30 mmHg and the chronic response pushes Pa_{CO_2} into the teens). At first glance the magnitude of this time-dependent acclimatization effect may seem rather miniscule, i.e., about 25% or less than a 2–3 l/min increase in $\dot{V}E$ over 10 days. Certainly this change may prove elusive to quantitate from day to day, but it is hardly insignificant in many other terms: (a) this acclimatization effect persists throughout one's stay in hypoxia; the cumulative effect is a substantial one, amounting to almost 3000 l/day, or over 1 million l/year; (b) it is *the* critical short-term adaptive mechanism for protection of arterial O_2 content in the human sojourner—especially during exercise; and (c) the important regulatory point is that $\dot{V}E$ ·is actually increasing over time, while measurable ventilatory chemical stimuli are changing in a manner which would otherwise *depress* ventilatory drive; i.e., Pa_{O_2} continually increases, and arterial [H$^+$] falls and then reaches a plateau at an alkaline level.

The principal questions here are (1) what causes the time-dependent hyperventilation over these many days in hypoxia, and (2) why does

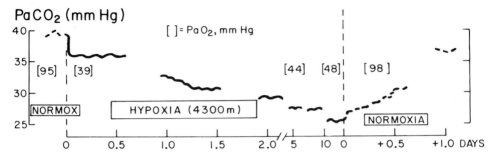

FIGURE 2.1. Ventilatory acclimatization to chronic hypoxia in the sea-level native sojourning at an altitude of 4300 m. The two essential features of acclimatization are (a) the time-dependent intensification of the level of hyperventilation between acute and chronic hypoxia and (b) the continued hyperventilation which gradually dissipates over time obtained upon acute restoration of normoxia. [Please note that the normoxic restoration data shown here were obtained after only 3 days of exposure to 4300 m (Dempsey et al. 1979); hyperventilation might still persist beyond 1 or 2 days' return to sea level following longer stays at high altitude.]

hyperventilation persist for some time when the hypoxic stimulus is withdrawn? Obviously the control system is "reset" so as to produce higher $\dot{V}_E$ at a lower Pa_{CO_2}—but this generalization doesn't help us to identify the missing stimulus which causes this resetting.

A Role for Bulk CSF [H⁺]?

The possibility which occurred to Severinghaus, Mitchell, and colleagues (1963) in the early 1960s was the concept of a "resetting" of cerebral fluid [H⁺] over time in hypoxia. Thus, a hypocapnic, alkaline CSF would partly counteract the increased carotid chemoreceptor activity and ventilatory drive in *acute hypoxia*, but over time of hypoxic exposure a quickly normalized CSF [H⁺] would no longer oppose peripheral chemoreceptor drive and thus $\dot{V}_E$ would rise further. Upon acute restoration of normoxia, P_{CO_2} would rise slightly (because of withdrawal of the peripheral chemoreceptor stimulus) and quickly acidify the CSF (already at a reduced [HCO₃⁻], thus providing the stimulus for continued hyperventilation in the absence of hypoxia.

This "unified control" theory provided a solid framework for the appreciation of chemoreceptor interaction and provided a substantial impetus for much research into the regulation of CSF ionic composition—but we do not think this concept explains venilatory acclimatization. In fact, it seems doubtful that cerebral fluid [H⁺] itself or perhaps even medullary chemoreception plays any stimulatory role in ventilatory acclimatization—at least as we currently understand the operation of this chemosensor.

The case against any role for [H$^+$] in the bulk CSF has been made previously (Dempsey and Forster 1982; Dempsey et al. 1979). In essence, contrary to initial postulates, we found that CSF [H$^+$] was compensated in an identical manner to arterial plasma during hypoxic exposure (or respiratory alkalosis). As hyperventilation proceeds with time in hypoxia, CSF [H$^+$] is reduced; and upon restoration of normoxia as ventilation falls and Pa$_{CO_2}$ rises slowly toward sea-level control values, CSF [H$^+$] increases.[1] Obviously neither change is positively correlated with the corresponding ventilatory response; in fact, ventilatory changes seem to occur *despite* the change in CSF [H$^+$].

A Role for Cerebral ISF [H$^+$]?

Clearly these data do *not* preclude a role for medullary chemoreception in ventilatory acclimatization if the [H$^+$] in the cerebral fluid environment of these receptors is regulated quite differently than it is in bulk CSF. This is precisely what Fencl et al. (1979) suggested based on their data in chronically hypoxic goats; Fencl et al. used ventricular cisternal perfusion techniques to determine transependymal ionic fluxes in order to estimate that the "deeper" cerebral interstitial fluid [HCO$_3^-$] was substantially lower and therefore [H$^+$] greater than in bulk CSF—presumably due to augmented lactic acid production in cerebral tissue. We confirmed this idea in rats by showing that concentrations of metabolic acids in cerebral tissue were indeed elevated in chronic hypoxia (see below). However, the problem remains of establishing whether a positive correlation exists over time between cerebral fluid [H$^+$] and the dynamic process of ventilatory acclimatization to hypoxia.

The near-ideal test of this possibility would be to determine the time course of change in ionic composition of the fluid in direct contact with the medullary chemoreceptors; but as their exact location remains unknown, we used two indirect approaches. First, in humans, we attempted to manipulate the cerebral interstitial fluid (ISF) [H$^+$] during acclimatization by causing and maintaining a metabolic alkalosis in the plasma (Dempsey et al. 1978), thus creating abnormal ionic gradients

---

[1] If one examines the [H$^+$] changes from chronic hypoxia over a 24-hr period of restoration of normoxia, clearly there are times during this period when continued hyperventilation (i.e., Pa$_{CO_2}$ < sea-level control) was present coincident with the presence of a CSF [H$^+$] which was significantly increased relative to sea-level control. Looking at these single points in time, one would conclude—falsely—that the stimulus for the continued hyperventilation was increased [H$^+$]; but once again if the correlation of the two variables over time is examined, the data clearly establish that ventilation is *decreasing* as [H$^+$] is increasing.

between plasma and bulk CSF which should alter the [H⁺] of the ISF and presumably that of the medullary chemoreceptor environment (Pappenheimer et al. 1965). The effects of chloride depletion alkalosis—achieved by thiazide diuretics and an NaCl-restricted diet—are shown in Table 2.1. Chloride depletion caused and maintained a moderately elevated plasma bicarbonate and depressed [Cl⁻]—relative to the control group—both at sea level (+5 meq/l [HCO_3^-], −10 meq/l [Cl⁻]), and throughout 6 days acclimatization to an altitude of 3200 m (+6 meq/l [HCO_3^-] and −13 meq/l [Cl⁻]). During chloride depletion alkalosis in plasma, CSF [HCO_3^-] and [Cl⁻] were reduced in a normal fashion during acclimatization[2]; this meant that the Cl⁻· HCO_3^-· and H⁺ gradients between plasma and CSF were elevated 1.5–7-fold over those in the control sojourner. Despite these manipulations, ventilatory acclimatization, measured at many time points during the 6 days at 3200 m and during acute return to normoxia, was unaltered from the control state (see the mean Pa_{CO_2} values in Table 1). Replacing chloride and stopping the diuretics during the fifth through ninth day of sojourn also had no significant effect on ventilatory acclimatization (not shown—see Dempsey et al. 1978). Similar results were obtained with $NaHCO_3$ loading during acclimatization.

A second approach to the question of acclimatization and its relation to ISF [H⁺] was to explore the source of the postulated cerebral metabolic

---

[2] It is of interest that CSF [HCO_3^-] and [H⁺] were regulated in a completely normal fashion; i.e., CSF [HCO_3^-] was reduced and [Cl⁻] rose the same amount for a given level of cerebral hypocapnia in the control group and in the group where plasma chloride was markedly depleted and plasma [HCO_3^-] was maintained at a constant normal value (sea level). The key regulator here appears to be a changing cerebral Pco_2 (Dempsey et al. 1979), which affects the rate of exchange of chloride between the cerebral tissue and cerebral extracellular fluid.

TABLE 2.1. Chloride depletion and ventilatory acclimatization.

	Pa_{CO_2} (mmHg)	[Cl⁻] (meq/l)			[HCO_3^-] (meq/l)		
		Pl	CSF	(Δ)	Pl	CSF	(Δ) *
Control							
sea level	40	109	126	(17)	24	25	(1)
+ 6 days, 3200 m	33	114	131	(17)	21	22	(1)
Thiazide							
sea level	42	99	118	(19)	29	28	(1)
+ 6 days, 3200 m	34	101	127	(26)	27	22	(5)

* All values are means; N = 6–8 humans per group.
† (Δ) refers to the difference between large-cavity CSF and arterial plasma. Note that during sojourn at 3200 m, in the hydrochlorothiazide plus NaCl-restricted group ("Thiazide"), CSF [Cl⁻] rose and CSF [HCO_3^-] was reduced similar to the control group, even though plasma concentrations were maintained at elevated [HCO_3^-] and reduced [Cl⁻]; thus CSF-plasma "gradients" for Cl⁻; HCO_3^-; and H⁺ were markedly elevated.

acidosis in brain stem and cortex in the rat—an animal which showed a humanlike ventilatory acclimatization to chronic hypoxia (Musch et al. 1983; Olson and Dempsey 1978). We did observe (see Fig. 2.2) an increased level of metabolic acids, mainly lactate and pyruvate, in cerebral tissue in chronic hypoxia at a time when $\dot{V}$E was high, and in turn these changes in brain tissue acids were shown to be secondary to the effects of both reduced arterial O$_2$ content and cerebral hypocapnia. Furthermore, even upon acute restoration of normoxia the persistent hyperventilation was accompanied by a slight but significantly elevated brain tissue lactate concentration (Table 2.2). Even intracellular [H$^+$] shifted in an acid direction during the acute stages of hypoxia. The problem is that the correlative data *over time* do not support a positive mediator role in ventilatory acclimatization for these changes in cerebral metabolic acids: (*a*) Note in Fig. 2.2 that as ventilatory acclimatization proceeds from *acute to chronic* hypoxia (note increased $\dot{V}$A : $\dot{V}$CO$_2$), metabolic acid levels are actually *reduced* in all extracellular and intracellular fluid compartments. (*b*) Upon acute restoration of normoxia the elevated metabolic acids in brain stem and cortex were attributable to the prevailing hypocapnia, because when Pa$_{CO_2}$ was increased to normal (via increased FICO$_2$), tissue lactate was restored to normal. Since the lactate was increased *because of* the continued hyperventilation, it is, therefore, difficult to postulate that the increased lactate was the *cause* of continued hyperventilation.

A third series of experiments to test a role for ISF [H$^+$] used chronic carotid body denervation so that cerebral fluid metabolic acidosis became the *only* available (known) chemical stimulus during hypoxic acclimatization (Smith et al. 1986). Previous data on this question were controversial (e.g., Forster et al. 1981; Steinbrook et al. 1983). We incorporated the following features into our study of goats: (*a*) We used multiple criteria to ensure completeness of carotid body denervation (CBX) and monitored plasma electrolytes to ensure their being normal. (*b*) We exposed intact and CBX animals to identical levels of arterial hypoxemia by using differing levels of hypobaria. (*c*) We used two levels and durations of hypoxia to test the consistency of the observations, including a "severe" level of hypoxemia (Pa$_{O_2}$ 28–34 mmHg), which according to our rat studies (see above) would ensure high levels of cerebral metabolic acid production throughout the sojourn and even cerebral *intra*cellular acidosis in acute hypoxia. We think the data—as summarized in Fig. 2.3—clearly support the necessity for peripheral chemoreception to ensure any significant level of ventilatory acclimatization: (*a*) CBX animals showed no significant hyperventilation throughout the period of hypoxic exposures; and (*b*) upon acute restoration of normoxia, CBX animals experienced hyperventilation which increased with time in normoxia, whereas the intact controls reduced their level of hyperventilation and ventilated less and less as the duration of normoxia was prolonged (up to 2 hr). The

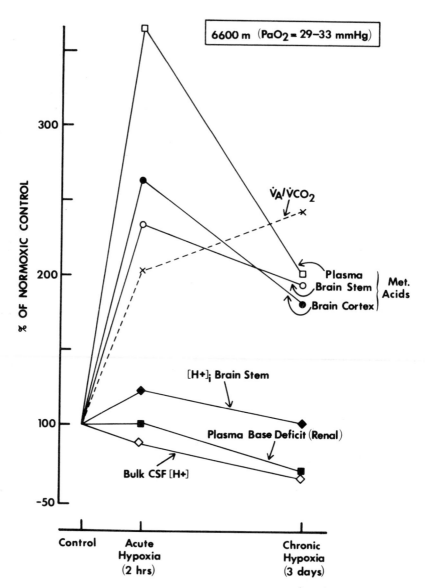

FIGURE 2.2. Changes in the level of hyperventilation ($\dot{V}_A : \dot{V}_{CO_2}$) vs. various indices of acid-base status in arterial plasma, in brain stem and cortex (as potential sources of cerebral ISF acidosis), and in bulk CSF (estimated) during acclimatization to severe chronic hypoxia in the rat. Mean values in *acute* hypoxia: Pa_{O_2} = 29 mmHg, Pa_{CO_2} = 19.8 mmHg, pHa = 7.59, and base excess = −0.8 meq/l; and in *chronic* hypoxia: Pa_{O_2} = 33 mmHg, Pa_{CO_2} = 16.5 mmHg, pHa = 7.49, and base excess = −7.2 mEq/l. (Musch et al. 1983.)

TABLE 2.2. Relationship of cerebral metabolic acids to ventilatory acclimatization to and deacclimatization from 6600 m altitude.*

	Pa$_{CO_2}$ (mm Hg)	Brain stem metabolic acids (mM/kg apt wt.)
Control normoxia	40	2.5
Hypoxia		
(Pa$_{O_2}$ 24–33 mmHg)		
Acute (2 hours)	21	7.0 †
Chronic (3 days)	17	4.7 †
Acute Restoration		
of normoxia		
(Pa$_{O_2}$ 95–110 mmHg)		
Hypocapnic	25	3.6
(ambient air)		
Normocapnic	40	2.5
(+ F$_I$CO$_2$)		

* All values are means; N = 6 to 18 rats per group.
† Mean value is different from normoxic control ($P < 0.05$).

only hint of at least some level of hyperventilation which might be attributable to stimuli other than peripheral chemoreception was the finding (Fig. 2.3) that the mean Pa$_{CO_2}$ tended to be visibly reduced (2–3 mmHg) in CBX animals exposed to severe vs. moderate hypoxemia (even though this difference was inconsistent and not significantly different from control in either condition of hypoxic exposure).

Ventilatory Response is NOT a Single Function of Cerebral Fluid [H$^+$]!

A major implication of the studies discussed above is that there is no apparent ventilatory response to marked changes in brain extracellular fluid [H$^+$]—especially that induced by cerebral metabolic acid production in the hypoxic subject. One explanation is that cerebral hypoxia—especially chronic hypoxia—may depress or inhibit this response to [H$^+$], but this clearly isn't the case with some types of altered [H$^+$]. For example, acute changes in P$_{CO_2}$ have profound effects on ventilation during acclimatization: (a) Certainly the reduced P$_{CO_2}$ upon exposure to acute hypoxia has marked inhibitory effects on the net ventilatory response. (b) The conventional hyperoxic CO$_2$ response slope is unchanged or increased in the sojourner at high altitude (Forster et al. 1971). (c) And during sleep in the sojourner at high altitude we recently observed that the characteristic periodic breathing was readily converted to regular rhythmic breathing and all apneas were eliminated with only 1 to 3 mmHg increments in Pa$_{CO_2}$ (achieved via increasing F$_I$CO$_2$)

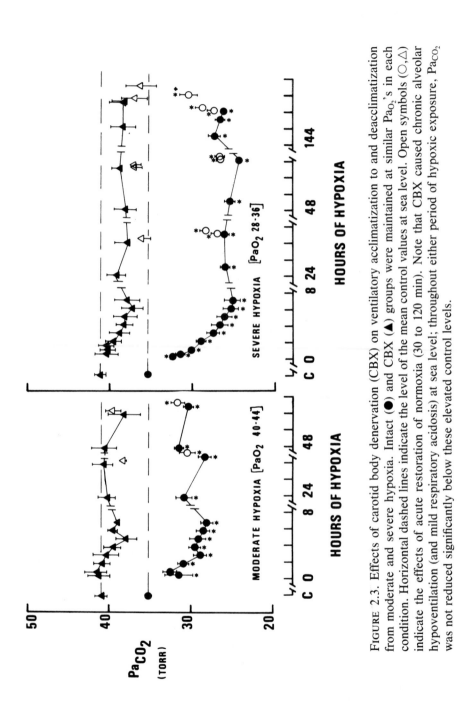

FIGURE 2.3. Effects of carotid body denervation (CBX) on ventilatory acclimatization to and deacclimatization from moderate and severe hypoxia. Intact (●) and CBX (▲) groups were maintained at similar Pa$_{O_2}$'s in each condition. Horizontal dashed lines indicate the level of the mean control values at sea level. Open symbols (○,△) indicate the effects of acute restoration of normoxia (30 to 120 min). Note that CBX caused chronic alveolar hypoventilation (and mild respiratory acidosis) at sea level; throughout either period of hypoxic exposure, Pa$_{CO_2}$ was not reduced significantly below these elevated control levels.

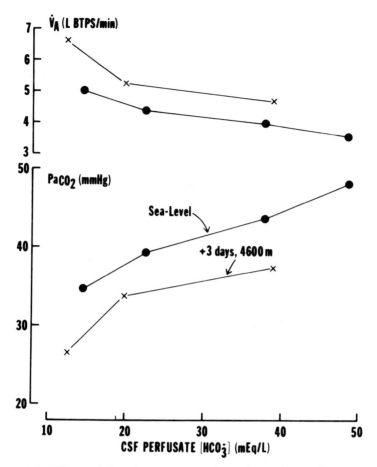

FIGURE 2.4. Effects of chronic exposure to hypoxia on the ventilatory response to changes in CSF [HCO$_3^-$] and [H$^+$] as induced by steady-state cisternal perfusion of artificial CSF in goats. The normal control values at each altitude were Pa$_{CO_2}$-41 mmHg, CSF [HCO$_3^-$]-23 meq/l, and est. CSF [H$^+$]-51 neq/l at sea level; and Pa$_{CO_2}$ = 33 mmHg, CSF [HCO$_3^-$]-20 meq/l, and est. CSF [H$^+$]-43 neq/l after 3 days at 4600 m. As evidenced by the near parallel response slopes at each altitude, chronic hypoxia had no systematic effect on the ventilatory response to acute, steady-state changes in CSF perfusate [H$^+$]. (For methods, see Jameson et al. 1983.)

(Berssenbrugge et al. 1983). Furthermore, we tested the ventilatory response of awake goats to acute steady-state changes in cerebral ECF [H$^+$] achieved via cisternal perfusion of mock CSF of varying [HCO$_3^-$] and [H$^+$], and we found it to be unchanged by chronic exposure to hypoxia (see Fig. 2.4).

We think the data are consistent with the postulate that the gain of the ventilatory response to different means of changing cerebral fluid [H$^+$]

may be markedly different. Acute studies in anesthetized animals using medullary surface pH measurements also showed a relative insensitivity of ventilatory responses to systemically induced metabolic changes in $[H^+]$ (Teppema et al. 1983) and hypoxia-induced metabolic acidosis (Kiwull-Schöne and Kiwull 1983)—as compared with similar changes in medullary $[H^+]$ induced by CO_2.

Why is there this difference in ventilatory responsiveness? One explanation is that the actual change in $[H^+]$ in the "real" environment (intracellular *or* transmembrane) of the medullary chemoreceptor is quite different in the different types of $[H^+]$ change. An alternative explanation for these differences in responsivity among the different forms of H^+ change is that $[H^+]$, at any site in the CNS, is only the apparent stimulus and that some related critical function or structure is the actual stimulus being defended by ventilatory responses. Reeves (1972) suggested that $[H^+]$ changes produce physiological effects (such as changes in enzyme activity) by changing the dissociation of imidazole groups (alpha-imidazole) located at critical positions on certain proteins, i.e., the "alpha-stat" hypothesis. Thus, just as Pappenheimer et al. (1965) originally proposed that ventilatory responses to hypercapnia and systemic or CSF perfusion-induced metabolic acidosis and alkalosis were a "single function" of estimated cerebral ISF $[H^+]$ in awake goats, others now claim [using temperature changes in air-breathing ectotherms (Hitzig 1982) or pharmacological modification of imidazole groups (Nattie 1985)] that ventilatory responses are uniquely related to the calculated ionization state of alpha-imidazole. This intriguing and fundamental concept needs more direct testing, but it might offer one explanation for the problem of apparent marked differences in how medullary chemoreceptors perceive changes of $[H^+]$ produced by different means.

The Primary Role for Peripheral Chemoreceptors?

The implication of these findings is that changes in carotid chemoreception probably play a major role in ventilatory acclimatization. On the other hand, the carotid body denervation studies alone do not provide a valid test of this possibility. Indeed, one cannot quantitate the contributions of a specific organ to any physiological phenomena by removing it. Bisgard and associates have provided a truly physiological test of the carotid body's contribution in the awake goat by isolating and perfusing the carotid body for many hours with blood of varying composition. This model allowed—for the first time—a separation of stimuli presented to carotid and medullary chemoreceptors in the unanesthetized animal (Bisgard et al. 1986a,b; Busch et al. 1985). The following findings to date in this model are of interest:

1. When carotid chemoreceptors are maintained hypoxemic and Po$_2$ is maintained normoxic in the systemic and cerebral vasculature, a normal time-dependent hyperventilation (and arterial hypocapnia) occurred over a 6-hr period and hyperventilation persisted upon acute restoration of normoxia. Thus, a hypoxic carotid body per se was sufficient to provide normal ventilatory acclimatization.
2. When carotid chemoreceptors were made hypoxemic and the systemic (and therefore cerebral) Pco$_2$ maintained normocapnic (via increased F$_I$CO$_2$), a normal time-dependent ventilatory acclimatization occurred over 4 hr. However, no continued hyperventilation occurred upon restoration of normoxia.
3. When the carotid body was made hypercapnic with a normal Po$_2$ background, V̇$_E$ increased acutely but showed no further time-dependent acclimatization over 6 hr.

How then could carotid bodies alone mediate time-dependent acclimatization? Either they may change their own output over time, *or* the increased afferent traffic may generate secondary time-dependent changes in the CNS. Millhorn et al. (1980) in anesthetized, paralyzed, deafferented cats used repeated sinus nerve electrical stimulation and demonstrated a "prolonged aftereffect" (following CSN stimulation) on phrenic nerve activity; this ventilatory aftereffect was eliminated when CNS serotonin was depleted. We doubt the applicability of these data to the physiological situation because: (1) increased CSN stimulation alone—via increased carotid body perfusate Pco$_2$—did not cause ventilatory acclimatization in the awake goat (Bisgard et al. 1986a.); and (2) depletion of CNS serotonin in awake rats did not affect any phase of ventilatory acclimatization to or deacclimatization from hypoxia (Olson, 1987). Furthermore, up to 70–80% pharmacological depletion of CNS (and carotid body) norepinephrine and dopamine also had no effect on ventilatory acclimatization in awake rats (McCrimmon 1983).

The output of the carotid sinus nerve may actually increase over time of hypoxic exposure. This would constitute a true change in carotid body "sensitivity" to hypoxia. Perhaps the chronically hypoxic carotid body responds to a progressive loss of dopamine-mediated inhibition (Bisgard et al. 1986b.; Olson et al. 1983). A changing cerebral acid-base status may also have some bearing on affecting a change in peripheral chemosensitivity, as shown with CSF perfusion experiments in awake goats (see Fig. 2.5). Note that as the steady-state CSF perfusate [HCO$_3^-$] was increased and CSF [H$^+$] was reduced, the ventilatory response to specific, transient stimulation of the peripheral chemoreceptors was augmented; i.e., a negative interaction occurred between chemoreceptors. While it is of interest that alkalinity does occur in the bulk CSF during sojourn in hypoxia, the relevance of this evidence of chemoreceptor interaction must await knowledge of changes in the true ionic composition of the medullary chemoreceptor environment during acclimatization.

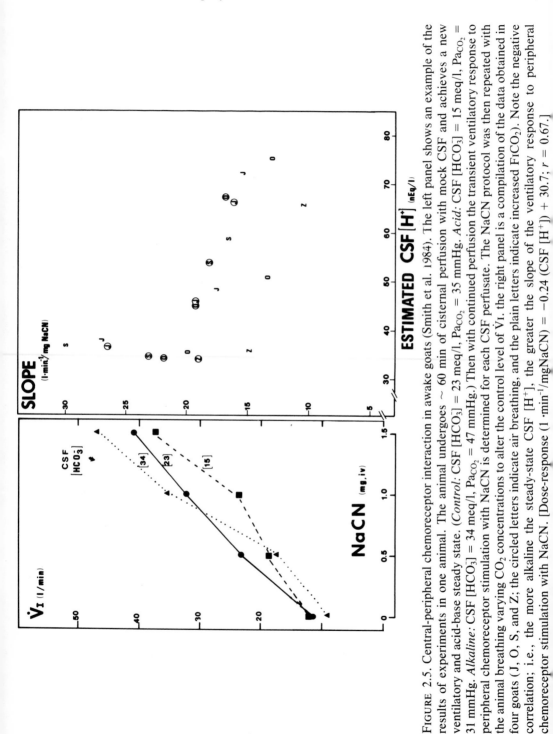

FIGURE 2.5. Central-peripheral chemoreceptor interaction in awake goats (Smith et al. 1984). The left panel shows an example of the results of experiments in one animal. The animal undergoes ~ 60 min of cisternal perfusion with mock CSF and achieves a new ventilatory and acid-base steady state. (*Control:* CSF [HCO$_3$] = 23 meq/l, Pa$_{CO_2}$ = 35 mmHg. *Acid:* CSF [HCO$_3$] = 15 meq/l, Pa$_{CO_2}$ = 31 mmHg. *Alkaline:* CSF [HCO$_3$] = 34 meq/l, Pa$_{CO_2}$ = 47 mmHg.) Then with continued perfusion the transient ventilatory response to peripheral chemoreceptor stimulation with NaCN is determined for each CSF perfusate. The NaCN protocol was then repeated with the animal breathing varying CO$_2$ concentrations to alter the control level of V̇I. the right panel is a compilation of the data obtained in four goats (J, O, S, and Z; the circled letters indicate air breathing, and the plain letters indicate increased FICO$_2$). Note the negative correlation; i.e., the more alkaline the steady-state CSF [H$^+$], the greater the slope of the ventilatory response to peripheral chemoreceptor stimulation with NaCN. [Dose-response (1·min^{-1}/mgNaCN) = −0.24 (CSF [H$^+$]) + 30.7; r = 0.67.]

Ventilatory "Deacclimatization" as a Separate Phenomenon?

Another fascinating implication of these data is that the process of time-dependent ventilatory acclimatization and the continued hyperventilation following restoration of normoxia may be mediated by two entirely different mechanisms. Bisgard's data strongly suggest that the continued hyperventilation does not occur unless systemic and cerebral hypocapnia accompany the hypoxemia administered to the carotid body. Clearly, cerebral hypocapnia *by itself* causes a substantial and prolonged "aftereffect," as shown by the many hours of continued spontaneous hyperventilation which was shown to *follow* 26 hr of maintained moderate, voluntary hyperventilation (Pa$_{CO_2}$ ~ 30 mmHg). That hypoxemia may also contribute significantly to this aftereffect was shown by the greater level of continued spontaneous hyperventilation which followed prolonged voluntary hyperventilation when a moderately hypoxemic background was present (Pa$_{CO_2}$ ~ 30 mmHg, Pa$_{O_2}$ ~ 50–55 mmHg); but this added effect of a hypoxemic background constituted only a small fraction of the total effect shown with (normoxic) hypocapnia alone on the continued spontaneous hyperventilation (Dempsey et al. 1975).

We cannot attribute this effect of prolonged cerebral hypocapnia on residual hyperventilation to cerebral acid-base status: (*a*) in bulk CSF, alkalinity prevailed at the termination of prolonged hypocapnia—and to the same extent in both normoxic and hypoxic backgrounds; and (*b*) as shown for acclimatized rats in Table 2, cerebral metabolic acid production seemed to follow (and be determined by) rather than precede (and cause) any continued hyperventilation upon acute restoration of normoxia. Perhaps prolonged cerebral hypocapnia might have an effect on some other as yet undefined CNS "stimulus" or some critical aspect of CNS neurotransmitter turnover. Considering the broad implications to the aftereffects of many conditions of chronic hyperventilation, and the substantial magnitude and duration of this response, the mechanisms of these aftereffects deserve further study.

Summary

Some rather dramatic changes in perspective have occurred over the past 20 years of research related to medullary chemoreceptors, chemoreceptor interactions, and their role in ventilatory acclimatization to hypoxia. Hallmarks of the early work included suggestions of a near-perfect regulation of the ionic composition of the ECF environment of the medullary chemoreceptor, additive effects of peripheral and central chemoreception on ventilatory output, and the view of alveolar ventilation as a single function of changes in chemoreceptor ISF [H^+]. Today virtually none of these ideas remains unchallenged. Indeed, some researchers doubt that medullary chemoreceptors even exist as distinct, unique, neuroanatomical entities; others question whether [H^+] is even a primary object of homeostatic control—including ventilatory control; some of us contend that ventilatory acclimatization proceeds either in spite of or indifferently to changes in cerebral fluid [H^+]. Hermann Rahn played a major role throughout the history of this continuing search for the mysterious mechanism of chemoreceptor "resetting." He provided the quantitative basis for originally defining the problem of acclimatization and in more recent times has been an enthusiastic and able proponent of the alpha-stat hypothesis. Dr. Rahn, you have—by example—provided the essential impetus to progress in yet another field of science. Many thanks!

Acknowledgments. This work was supported in part by a grant from the NHLBI and the USAMRDC. C.A. Smith is a Parker B. Francis Fellow in Pulmonary Research. We wish to thank Ms. Pamela Hamm for her excellent preparation of the manuscript and to acknowledge the contributions from many of our collaborators whose original work is cited in this review.

REFERENCES

Berssenbrugge AD, Dempsey JA, Iber C, Skatrud JB, Wilson P (1983). Mechanisms of hypoxia-induced periodic breathing during sleep in humans. *J Physiol* (*London*) 343: 507–524

Berssenbrugge A, Dempsey JA, Skatrud JB (1984). Effects of sleep state on ventilatory acclimatization to chronic hypoxia. *J. Appl Physiol* 57: 1089–1096

Bisgard GE, Busch MA, Daristotle L, Forster HV (1986). Carotid body hypercapnia does not elicit ventilatory acclimatization in goats. *Respir. Physiol* 65: 113–125

Bisgard GE, Busch MA, Forster HV (1986). Ventilatory acclimatization to hypoxia is not dependent upon cerebral hypocapnic alkalosis. *J Appl Physiol* 60: 1011–1015

Busch MA, Bisgard GE, Forster HV (1985). Ventilatory acclimatization to hypoxia is not dependent on arterial hypoxemia. *J Appl Physiol* 58: 1874–1880

Dempsey JA, Forster HV (1982). Mediation of ventilatory adaptations. *Physiol Rev* 62: 262–346

Dempsey JA, Forster HV, Bisgard GE, Chosy LW, Hanson PG, Kiorpes AL, Pelligrino DA (1979). Role of cerebrospinal fluid [H^+] in ventilatory deacclimatization from chronic hypoxia. *J Clin Invest* 64: 199–205.

Dempsey JA, Forster HV, Chosy LW, Hanson PG, Reddan WG (1978). Regulation of CSF [HCO_3] during long-term hypoxic hypocapnia in man. *J Appl Physiol: Respirat Environ Exercise Physiol* 44: 175–182

Dempsey JA, Forster HV, Gledhill N, Do Pico GA (1975). Effects of moderate hypoxemia and hypocapnia on CSF [H$^+$] and ventilation in man. *J Appl Physiol* 38: 665–674

Fencl VR, Gabel A, Wolfe D (1979). Composition of cerebral fluids in goats adapted to high altitude. *J Appl Physiol: Respirat Environ Exercise Physiol* 47: 508–513

Forster HV, Bisgard GE, Klein JP (1981). Effect of peripheral chemoreceptor denervation on acclimatization of goats during hypoxia. *J Appl Physiol: Respirat Environ Exercise Physiol* 50: 392–398

Forster HV, Dempsey JA, Birnbaum ML, Reddan WG, Thoden J, Grover RF, Rankin J (1971). Effect of chronic exposure to hypoxia on ventilatory response to CO$_2$ and hypoxia. *J Appl Physiol* 31: 586–592

Hitzig BM (1982). Temperature-induced changes in turtle CSF pH and central chemical control of ventilation. *Respir Physiol* 49: 205–222

Jameson LC, Smith CA, Dempsey JA (1983). A method for cisterna magna perfusion of synthetic CSF in the awake goat. *J Appl Physiol* 55: 1623–1629

Kiwull-Schöne J, Kiwull P (1983). Hypoxia modulation of central chemosensitivity. In: Schläfke ME, Koepchen HP, See WR (eds) *Central Neurone Environment*. Springer-Verlag, Berlin Heidelberg

McCrimmon D, Dempsey JA, Olson EB Jr (1983). Effect of catecholamine depletion on ventilatory control in unanesthetized normoxic and hypoxic rats. *J Appl Physiol* 55: 522–528

Millhorn DE, Eldridge FL, Waldrop TG (1980). Prolonged stimulation of respiration by endogenous central serotonin. *Respir Physiol* 42: 171–188

Musch TI, Dempsey JA, Smith C, Mitchell G, Bateman NT (1983). Metabolic acids and [H$^+$] regulation in brain tissues during acclimatization to chronic hypoxia. *J Appl Physiol* 55: 1486–1495

Nattie EE (1985). Intracisternal diethylpyrocarbonate (an imidazole binding agent) inhibits rabbit central chemosensitivity. *Fed Proc* 45: 428 (abstract)

Olson EB Jr, Dempsey JA (1978). Rat as a model for human-like ventilatory adaptation to chronic hypoxia. *J Appl Physiol: Respirat Environ Exercise Physiol* 44: 763–769

Olson EB Jr, Vidruk EH, McCrimmon DR, Dempsey JA (1983). Monoamine neurotransmitter metabolism during acclimatization to chronic hypoxia. *Respir Physiol* 54: 79–96

Olson, EB Jr. Ventilatory adaptation to hypoxia occurs in serotonin-depleted rats. *Respir. Physiol.* 69: 227–235, 1987.

Pappenheimer JR, Fencl V, Heisey SR, Held R (1965). Role of cerebral fluids in control of respiration as studied in unanesthetized goats. *Am J Physiol* 208: 436–450

Rahn H, Otis AB (1949). Man's respiratory response during and after acclimatization to high altitude. *Am J Physiol* 157: 445–462

Reeves RB (1972). An imidazole alphastat hypothesis for vertebrate acid-base regulation: Tissue carbon dioxide content and body temperature in bullfrogs. *Respir Physiol* 14: 219–236

Severinghaus JW, Mitchell RA, Richardson BW, Singer MM (1963). Respiratory control at high altitude suggesting active transport regulation of CSF pH. *J Appl Physiol* 18: 1155–1166

Smith CA, Bisgard GE, Nielsen AM, Daristotle L, Kressin NA, Forster HV, Dempsey JA (1986). Carotid bodies are required for ventilatory acclimatization to chronic hypoxia. *J Appl Physiol* 60: 1003–1010

Smith CA, Jameson LC, Mitchell GS, Musch TI, Dempsey JA (1984). Central-peripheral chemoreceptor interaction in awake cerebrospinal fluid-perfused goats. *J Appl Physiol* 56: 1541–1549

Steinbrook RA, Donovan JC, Gabel RA, Leith DE, Fencl V (1983). Acclimatization to high altitude in goats with ablated carotid bodies. *J Appl Physiol* 55: 16–21

Teppema LJ, Barts PW, Folgering HT, Evers JAM (1983). Effects of respiratory and (isocapnic) metabolic arterial acid-base disturbances on medullary extracellular fluid pH and ventilation in cats. *Respir Physiol* 53: 379–395

3

Adventitial and Medial Proliferation of Lung Vessels in Neonatal Pulmonary Hypertension: The Calf at High Altitude

K.R. Stenmark, J. Fasules, A. Tucker, and J.T. Reeves

In human newborns, pulmonary hypertension is a serious problem. Some form of pulmonary hypertension has been estimated to occur in the newborn once per 1454 live births in those without evidence of pulmonary aspiration, and a higher incidence is estimated for newborns with aspiration syndromes (Goetzman and Reimenschneider 1980). The mortality in such neonatal pulmonary hypertension has been estimated to vary from 20 to 50% (Hoffman and Heyman 1984). Those infants who survive often have continuing respiratory morbidity which could be caused by barotrauma from chronic high-pressure mechanical ventilation using high oxygen mixtures (Wung et al. 1985). That the causative mechanisms are not clear is suggested by recent reviews and reports noting the wide variety of conditions associated with the pulmonary hypertension that include postmaturity, group B streptococcal pneumonia, meconium aspiration, asphyxia, pulmonary hypoplasia, and disorders of surfactant production (Fox and Durara 1985; Hallman and Kanakaanpaa 1980; James et al. 1984; Perkin and Anas 1984).

One question to be assessed is whether, as the name implies, in persistent pulmonary hypertension of the newborn, the disorder necessarily results from some abnormality in utero, currently the prevailing view (Geggel and Reid 1984; Haworth and Reid 1976). In support of this view, the pulmonary vascular resistance falls progressively near the end of fetal life coincident with a large increase in new vessel growth. Any process interfering with new vessel growth would set the stage for pulmonary hypertension after birth (Levin et al. 1976). Further, the primary evidence favoring a prenatal abnormality is the severe and widespread nature of the morphological changes in the lung vessels, occurring within a postnatal life span lasting only a few hours or a few days (Murphy et al. 1981). In rats made hypoxic, 10 to 14 days were required to induce vascular changes in the lung of a magnitude comparable to those seen in the infants (Meyrick and Reid 1978; Rabinovitch et al. 1979). In addition, prenatal vascular changes have been produced in fetal lambs by ligating the ductus arteriosus in utero 36 days prior to sacrifice of the fetus. The changes included considerable increase of collagen in the media and adventitia (Ruiz et al. 1972), as has been reported in the human condition (Geggel and Reid 1984).

Alternatively, it is possible that some neonatal pulmonary hypertension arises after birth, without an obligatory prenatal abnormality. Murphy et al. noted the surprising finding that infants dying after meconium aspiration had morphological changes identical to those of persistent pulmonary hypertension of the newborn (Murphy et al. 1984). As they felt the vascular changes preceded birth, they presumed a link between the abnormal lung vessels and meconium aspiration. The alternative hypothesis, that the vascular changes could develop remarkably fast after birth, was also raised. In this regard, fatal pulmnary hypertension of the newborn has occurred where the onset of symptoms was not noted until after the immediate neonatal period (Burnell et al. 1972). Thus, signs of pulmonary hypertension may be detected after a latent period of minutes to days, during which time the infant is considered healthy.

Our interest in the problem was stimulated by the following patient: JE, a 3460-g, term, appropriate-for-gestational-age infant, was born, following an uncomplicated pregnancy, to a 19-year-old mother, who was gravida 2, para 2. The fetal membranes were ruptured 1 hr before delivery, and the amniotic fluid was slightly stained with meconium. A sterile, spontaneous vaginal delivery was accomplished without complication. The APGAR scores of 8 at 1 min and 9 at 5 min were consistent with a vigorous infant. The child remained well, with good color and no respiratory distress, until 29 hr of age, when cyanosis and difficult respirations were noted. The arterial oxygen tension (Pa_{O_2}) was found to be 9 mmHg. Following the placement of an endotracheal tube, the child was ventilated with 100% oxygen; however, the Pa_{O_2} remained below 30 mmHg. An echocardiogram revealed normal cardiac anatomy: However, prolonged right ventricular systolic time intervals (preejection/ejection periods) and the presence of midsystolic closure of the pulmonary valve were considered to reflect severe pulmonary hypertension. The pulmonary hypertension plus the hypoxemia despite oxygen administration suggested right-to-left intracardiac shunting of blood. Tolazoline given intravenously to reduce the pulmonary hypertension and hypoxemia failed to increase the Pa_{O_2}. Despite intensive medical therapy, including continued oxygen and mechanical ventilation, the infant died on the fourth day of life. The findings at autopsy included severe pulmonary congestion, edema and abundant intra-alveolar squamous debri and alveolar macrophages. There was focal intra-alveolar and interstitial hemorrhage. No bacterial pathogens were noted, and the culture was negative. On microscopic examination there was striking adventitial fibrosis, medial hypertrophy, and thickened elastic lamina (Fig. 3.1). The heart showed right ventricular hypertrophy and dilatation, and there was a probe patent foramen ovale as well as a patent ductus arteriosus.

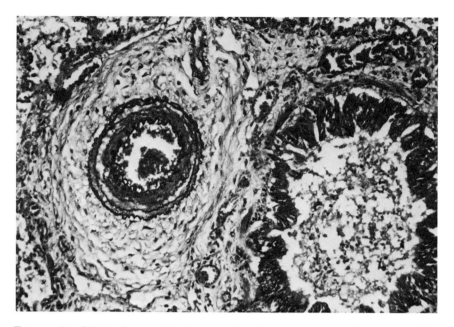

FIGURE 3.1. Photomicrograph of a muscular pulmonary artery and accompanying airway in a 4-day-old infant dying of persistent pulmonary hypertension. The vessel shown illustrates the medial hypertrophy and marked adventitial thickening. Also noted at autopsy was right ventricular hypertrophy and patency of both the ductus arteriosis and foramen ovale, findings consistent with the clinical course of right-to-left shunting of blood and severe hypoxemia noted in this infant. (Pentachrome stain, 500×.)

We were impressed by the apparently normal appearance of the infant for the first day of life, by the progressive nature of the illness, and, following autopsy, by the proliferative nature of the vascular lesions, in particular by the thickness of and apparent collagen deposition in the adventitia. We wondered whether the vascular morphology might have been normal or nearly normal at birth, with the marked abnormalities developing rapidly thereafter. An animal model was needed to test the hypothesis.

The requirements for the model were that the animal should be normal for a period after birth, and that upon application of an appropriate stimulus, pulmonary hypertension should rapidly develop. With time, pulmonary pressure should exceed systemic pressure, there should be a right-to-left shunt, and the hypertension should not be easily reversible with oxygen administration. Previous work with the calf suggested that were the newborn placed in a sufficiently hypoxic environment, the requirements for the model might be met (Reeves and Leathers 1967). The calf has a particularly reactive pulmonary circulation, and chronic hypoxia is known to be accompanied by rather marked hyperplasia of the media and adventitia of the lung vessels (Jaenke and Alexander 1973; Tucker et al. 1975). That hypoxia-induced morphological changes can occur quickly has been suggested by the work of Sobin et al., where by 8 hr of hypoxia there has been an increase in the number of fibroblasts in the arterial wall. By 24 hr the number had tripled (Sobin et al. 1983). That hypoxia is quickly followed by an increase in DNA synthesis in various lung cells has been shown (Niedenzu et al. 1981; Voelkel et al. 1977). That increased metabolic activity of the adventitia is both early and large has also been suggested (Meyrick and Reid 1979). That collagen deposition accompanies the vascular proliferative change of hypoxia is suggested by the finding that an inhibitor of collagen production inhibits the development of hypoxic pulmonary hypertension (Kerr et al. 1987). Thus, from the above we considered it worthwhile to investigate the proliferative nature of the vascular changes in the newborn calf at high altitude. If we could demonstrate the rapid development of vascular thickening, particularly involving the adventitia, in association with severe pulmonary hypertension, then the potential for postnatal development of human pulmonary hypertension should be considered.

Calves 1 to 2 days of age were placed in an altitude chamber which was evacuated to a pressure of 445 mmHg, equivalent to approximately 4300 m altitude, and constantly maintained there for 14 to 15 days. Control newborn calves were maintained at Fort Collins, Colorado, having a barometric pressure of 640 mmHg and an elevation of 1600 m. Hemodynamic measurements made in the unanesthetized calves at the end of the study (Fig. 3.2) showed that the high-altitude group had developed a mean pulmonary arterial pressure in excess of 100 mmHg, while the control calves had pressures less than 30 mmHg. The systemic arterial pressure was also lower in the high; than in the low; altitude calves. Thus, the calves at high altitude had developed severe pulmonary hypertension, with systemic hypotension.

The hypoxia of high altitude had, of course, reduced the oxygen saturation in the mixed venous and the arterial blood. However, there was, in the high-altitude calves, a progressive fall in the arterial oxygen tension from the first 2 to 4 days of exposure (Pa_{O_2}) = 36 mmHg) to the final measurement (Pa_{O_2} = 29 mmHg, $P<0.05$). The mixed venous oxygenation also decreased during the exposure. The decreased oxygenation was associated with the development of right-to-left shunting of blood (as determined by indicator dilution curves) primarily through a patent foramen ovale, but also through the ductus arteriosus. The administration of 100% oxygen to breathe via a muzzle mask for 5 to 10 min did not increase the pulmonary arterial pressure. That the pressure remained above that in the aorta was confirmed by direct measurement of pressure, by the persistence of a right-to-left shunt through the ductus arteriosus, and by the failure of the arterial oxygen tension to rise above

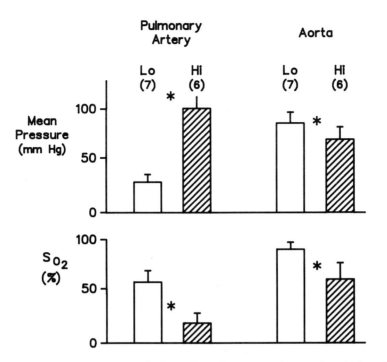

FIGURE 3.2. Measurements made from the pulmonary artery and aorta in calves age 14–22 days at 1800 m (PB=640 mmHg) and in calves age 16–18 days which had for 14–15 days lived at 4300 m (PB=445 mmHg). Shown are mean values and 1 standard deviation for the groups for mean pulmonary arterial and aortic pressure (*top*) and for oxygen saturation in mixed venous and arterial blood (*bottom*). The open bars represent the seven low-altitude calves, and the hatched bars represent the six high-altitude calves. * Indicates that high-altitude measurements differ (P<0.05) from low-altitude values.

100 mmHg. Subsequent studies showed an inability of prostacyclin and sodium nitroprusside (infused directly into the pulmonary artery) to lower pulmonary vascular resistance below systemic levels in newborn calves exposed to chronic hypoxia. (Orton et al. 1988) Thus a marked change in vascular reactivity accompanies the development of severe pulmonary hypertension in these newborn animals.

At the conclusion of the hemodynamic measurements, the calves were killed and the lungs examined histologically. The remarkable changes in the arterioles included not only medial hypertrophy, but also marked hyperplasia and thickening of the adventitia (Fig. 3.3). The changes were seen in all calves and were widespread within the sections examined. The lumina of the small arteries were reduced; the intima by light microscopy did not appear abnormal, but endothelial cell injury was evident by electron microscopy (Fig. 3.4*a*). The external elastic lamina appeared to

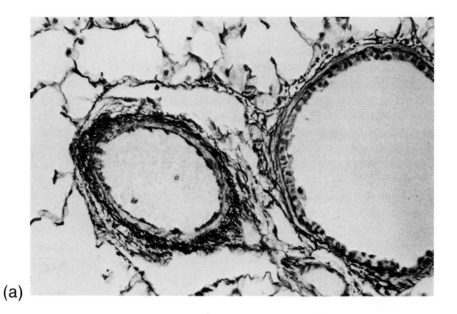

(a)

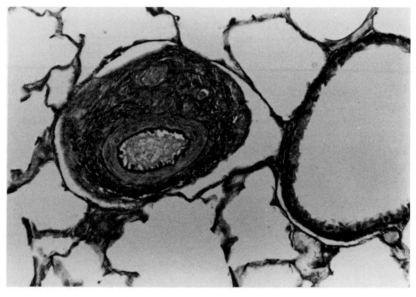

(b)

FIGURE 3.3. *a.* Photomicrograph of a muscular pulmonary artery and the accompanying airway in a 17-day-old calf born and living at 1500 m. The mean pulmonary artery pressure measured was 27 mmHg. (Luna stain, 500.)

b. Photomicrograph of a muscular pulmonary artery and accompanying airway in a 17-day-old calf after 15 days at 4300 m. Mean pulmonary artery pressure was 98 mmHg. Pulmonary artery pressure remained suprasystemic, with 100% oxygen breathing. Luminal narrowing, medial hypertrophy, and adventitial thickening are present. Note also the disruption of the external elastic.

be disrupted in many of the arteries. Within the adventitia there were sworls of what appeared to be fibrous tissue. Thus the orientation of the adventitial cells was distorted from the usual concentric pattern. In addition, both the trichrome and pentachrome stains suggested a large increase in the collagen content of the adventitia. The adventitia, by both light and electron microscopy, contained new vessel formation and the presence of extravesated erythrocytes and stimulated fibroblasts (Fig. 3.4b). Morphometrical analysis confirmed the marked increase in the medial and adventitial layers and the reduction in the lumen size in the high-altitude calves (Stenmark et al. 1987).

The physiological similarities between the newborn calf at altitude and the infant with pulmonary hypertension include development of severe pulmonary hypertension, the right-to-left shunting of blood, the increasing hypoxemia, and a decrease in pulmonary vasoreactivity. The morphological similarities included the marked reduction in lumen diameter, the thickened media, and the marked increase in the adventitia, with collagen deposition. However, there were obvious differences between the calf and the child, in particular, that the stimulus for the calf was hypoxia, while that for the child remains unknown. The 2-week duration of hypoxic exposure for the calf exceeded the 4-day life span of the child. The long-term reversibility of the hypertension in the calf was not examined. Also, the histological features seen in the calf differed in some respects from those in the child. However, the findings in the calf did indicate that in this animal model, striking vascular changes can be induced in a short time and these are associated with particularly severe pulmonary hypertension (Stenmark et al. 1987).

The present experiments in no way deny that prenatal influences are important in the development of pulmonary hypertension of the newborn. The results, however, do raise the possibility that in the newborn the capacity for rapid development of severe vascular changes exists. The results also emphasize previously published studies in infants which noted the potentially important role of the adventitia in the proliferative component in chronic pulmonary hypertension. Whether or not the proliferative changes contribute to the relative irreversibility of the pulmonary hypertension deserves further investigation.

When pulmonary vascular remodeling occurs in utero, and in addition there are postnatal influences which stimulate proliferative changes, these latter changes could contribute markedly to the clinical problem of pulmonary hypertension. Thus we see two possibilities which are not mutually exclusive: First, the abnormal fetal pulmonary vasculature persisting till birth may become rapidly more abnormal after birth with the subsequent development of proliferative changes. Second, even the normal pulmonary vascular bed at birth may, with certain stimuli, develop proliferative changes which lead to severe pulmonary hypertension. Because the mechanisms for the development of neonatal pul-

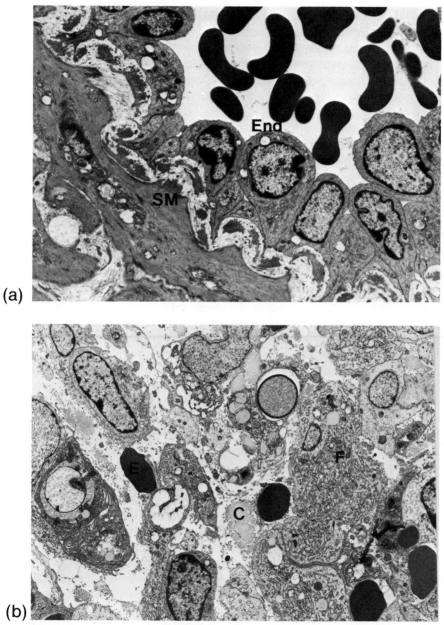

(a)

(b)

FIGURE 3.4. *a.* Electron micrograph of a small pulmonary artery ($\sim$ 120 μm) showing endothelial cells (End) with vacuoles and dense bodies. A smooth muscle cell (SM) is also shown outside the internal elastic lamina. (4600$\times$.)

b. Electron micrograph of the adventitia of the same pulmonary artery shown above. Note the presence of fibroblasts (F) with dense endoplasmic reticulum, collagen (C), and erythrocytes (E). (4600$\times$.)

monary hypertension remain unclear, the newborn calf at high altitude may provide a useful animal model for future investigation of etiology and treatment.

REFERENCES

Burnell RH, Joseph MC, Lees MH (1972). Progressive pulmonary hypertension in newborn infants. *Am Dis Child* 123: 167–170

Fox WW, Durara S (1985). Persistent pulmonary hypertension in the neonate: Diagnosis and treatment. *J Ped* 103: 505–514

Geggel RL, Reid LM (1984). The structural basis of PPHN. *Clin Perniatol* II: 525–549

Goetzman BW, Reimenschneider TA (1980). Persistence of the fetal circulation. *Ped Rev* 2: 37–40

Hallman M, Kanakaanpaa K (1980). Surfactant deficiency in persistent fetal circulation. *Euro J Ped* 134: 129–134

Haworth S, Reid L (1976). Persistent fetal circulation: Newly recognized structural features. *J Ped* 88: 614–620

Hoffman JIE, Heyman MA (1984). Persistent pulmonary hypertension syndromes in the newborn. In: Weir ED, Reeves JT (eds) *Pulmonary Hypertension.* Future Publishing Co., Mt. Kisco, NY, chap 2, pp 45–72

Jaenke RS, Alexander AF (1973). Fine structural alterations of bovine peripheral pulmonary arteries in hypoxia-induced hypertension. *Am J Path* 73: 377–390

James DK, Chiswick ML, Harkes A, Williams M, Hallworth J (1984). Non-specificity of surfactant deficiency in neonatal respiratory disorders. *Br Med J* 288: 1635–1638

Kerr JS, Ruppert CL, Sterbenz G, Frankel HM, Yu SY, Riley DJ (1987). Prevention of chronic hypoxic pulmonary hypertension in the rat by an inhibitor of collagen production inhibits the development of hypoxic pulmonary hypertension. *Am Rev Respir Dis* 135: 301–306

Levin DL, Rudolph AM, Heyman MA, Phibbs RH (1976). Morphologic development of the pulmonary vascular bed in fetal lambs. *Circ* 53: 144–151

Meyrick B, Reid L (1978). Effect of continued hypoxia on rat pulmonary arterial circulation. An ultrastructural study. *Lab Invest* 38: 188

Meyrick B, Reid L (1979). Hypoxia and incorporation of ^{3}H thymidine by cells of the rat pulmonary arteries and alveolar wall. *Am J Pathol* 96: 51–70

Murphy JD, Rabinovitch M, Goldstein JD, Reid LM (1981). The structural basis of persistent pulmonary hypertension of the newborn infant. *J Ped* 98: 962–967

Murphy JD, Vawter GF, Reid LM (1984). Pulmonary vascular disease in fatal meconium aspiration. *J Ped* 104: 758–762

Niedenzu C, Grasedyck K, Voelkel NF, Bittman S, Lindner J (1981). Proliferation of lung cells in chronically hypoxic rats. *Int Arch Occup Environ Hlth* 48: 185–193

Orton EC, Reeves JT, Stenmark KR, Pulmonary Vasodilation in Calves with structurally-altered Pulmonary Vessels and Pulmonary Hypertension. J. Appl. Physiol. (in press) Dec. 1988

Perkin RM, Anas NG (1984). Pulmonary hypertension in pediatric patients. *J Ped* 105: 511–522

Rabinovitch M, Gamble W, Nadas AS, Miettinen OS, Reid L (1979). Rat pulmonary circulation after chronic hypoxia: Hemodynamic and structural features. *Am J Physiol* 236: H818–H827

Reeves JT, Leathers JE (1967). Post natal development of pulmonary and bronchial arterial circulations in the calf and the effects of chronic hypoxia. *Anat Rec* 157: 641–655

Ruiz U, Piasecki GJ, Balogh K, Jackson BT (1972). An experimental model for fetal pulmonary hypertension. *Am J Surg* 123: 471

Sobin SS, Tremer HA, Hardy JD, Chiodi HP (1983). Changes in arteriole in acute and chronic hypoxic pulmonary hypertension and recovery in rat. *J Appl Physiol* 55: 1445–1455

Stenmark KR, Fasules JW, Voelkel NF, Henson J, Tucker A, Wilson H, Alexander AF, Reeves JT (1987). Severe pulmonary hypertension and arterial adventitial changes in newborn calves at 4300 m. *J Appl Physiol* 62: 821–830

Tucker A, McMurtry IF, Reeves JT, Alexander AF, Will DH, Grover RF (1975). Lung vascular smooth muscle as a determinant of pulmonary hypertension at high altitude. *Am J Physiol* 228(3): 762–767

Voelkel NF, Wiegers U, Sill V, Trautman J (1977). A kinetic study of lung DNA synthesis during simulated chronic high-altitude hypoxia. *Thorax* 32: 578–581

Wung JT, James LS, Kilhevsky E, James E (1985). Management of infants with severe respiratory failure and persistence of the fetal circulation, without hyperventilation. *Pediatrics* 76: 488–494

4

Muscle Function Impairment in Humans Acclimatized to Chronic Hypoxia

P. CERRETELLI, P.E. DI PRAMPERO, AND H. HOWALD

Introduction

The classical notion that the maximal aerobic performance ($\dot{V}O_2max$) of humans at altitude is essentially limited by the reduced convective and diffusive flow of oxygen to the tissues consequential to the decreased inspired O_2 pressure (PIO_2), and that acclimatization may partially compensate for this deficit by enhancing erythropoiesis, is the result of an oversimplification and may need substantial revision. Indeed, there are hints that chronic hypoxia may seriously impair the function of skeletal muscle so that the observed decrease in maximal aerobic power at altitude may be the consequence not only of a defect in O_2 transport but also of the metabolic power failure of the working tissues. Two direct and two indirect observations appear to be relevant (Cerretelli and di Prampero 1985): (1) a consistent reduction of muscle mass, particularly in the legs, which is a common experience of the Himalayan mountain climbers; (2) the relatively low arteriovenous O_2 difference (~ 10 ml/100 ml of blood) found during submaximal and maximal exercise at Mt. Everest base camp; (3) the similarity of the percent reduction of $\dot{V}O_2max$ with decreasing barometric pressure (PB) in acutely and chronically exposed individuals; and (4) the failure of acclimatized individuals to resume sea-level $\dot{V}O_2max$ upon sudden hyperbaric O_2 breathing.

1. The reduction of muscle mass has been reported, anecdotally, by many climbers. The only experimental measurements we are aware of are those carried out by Boutellier et al. (1983), who, 2 weeks after leaving altitude, found an average 12% decrease of muscle mass as assessed by computer tomography of the thigh in a group of climbers of the 1981 Swiss Mt. Lhotse expedition.
2. Mixed venous blood O_2 and CO_2 pressures and contents were determined at Mt. Everest base camp (5350 m) by Cerretelli (1976a) on five subjects by the N_2-CO_2 rebreathing technique. From the above data and simultaneous $\dot{V}O_2$ measurements, $\dot{Q}$ was calculated at rest (sitting) and during graded cycloergometric loads, the heaviest corresponding to $\sim85\%$ of individual $\dot{V}O_2max$. In the latter condition, the average $\dot{Q}$

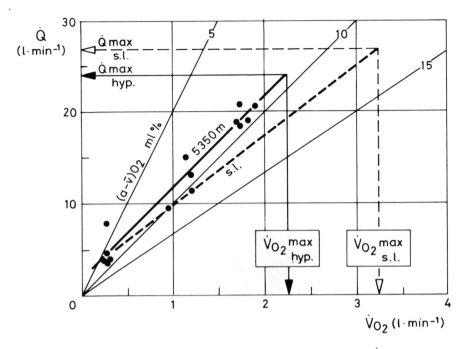

FIGURE 4.1. Cardiac output ($\dot{Q}$) as a function of O_2 consumption ($\dot{V}$). *Thick line:* Five subjects measured at 5350 m after 6–8 weeks' exposure to altitudes between 5350 and about 8000 m. *Dashed line:* Average for a group including the same subjects, at sea level. $\dot{V}_{O_2}$ max$_{hyp}$ and $\dot{V}_{O_2}$ max$_{s.l.}$ = average measured values. $\dot{Q}$max$_{hyp}$ and $\dot{Q}$max$_{s.l.}$ = average extrapolated values. iso-(a-$\bar{v}$)o_2 lines for 5,10,15 ml0$_2$ per 100 ml of blood are also drawn. (From Cerretelli 1976a, modified.)

was close to 20 l·min^{-1}, while the arteriovenous O_2 difference was ~ 9 ml/100 ml of blood. Assuming a 10% increase of $\dot{Q}$ with raising $\dot{V}_{O_2}$ to maximum, the difference between $\dot{Q}$max at sea level and at 5350 m would be of the order of 10–15% (see Fig. 4.1). Pugh (1964), after a sojourn at 5800 m of several months, found a drop of $\dot{Q}$max of 22%.

3. The average reduction of $\dot{V}_{O_2}$max as a function of P_{IO_2}, at least up to a P_B = 350 Torr (~ 6000 m), appears to be the same in acute and chronic hypoxia and is independent of blood hemoglobin concentration [Hb], as shown in Fig. 4.2. The average increase of [Hb] in the course of acclimatization may totally compensate for the decreased percent O_2 saturation (% So_2), so that Ca_{O_2} remains essentially unchanged. This is obviously not the case for acute hypoxia, where % So_2 at a simulated altitude of 5500 m is around 70%. Maximal cardiac output ($\dot{Q}$max), on the other hand, as indicated in Fig. 4.1, decreases only 10 to 20% in chronic hypoxia, with the drop being almost proportional to the reduction of maximal heart rate (HR max) (Cerretelli 1976a and 1976b;

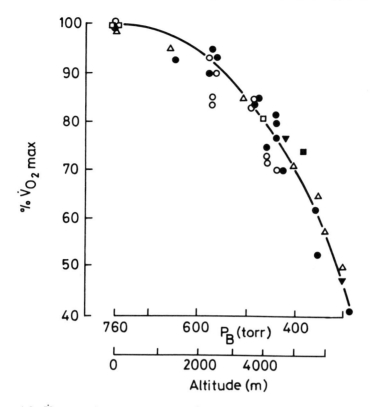

FIGURE 4.2. $\dot{V}_{O_2}$max (as a percentage of the sea-level value) as a function of barometric pressure (P_B) or altitude. $\triangle,\square,\bigcirc$ = acute hypoxia; $\blacktriangledown,\blacksquare,\bullet$ = chronic hypoxia; $\blacktriangledown$ = altitude natives ($n = 9$); $\blacksquare$ = average of 32 acclimatized lowlanders; $\bullet$ = acclimatized lowlanders; $\square$ = acute hypoxia in a decompression chamber ($n = 15$); $\triangle$ = acute hypoxia breathing hypoxic mixtures; $\bigcirc$ = newcomers to altitude. (Redrawn from Cerretelli and di Prampero 1985.)

Pugh 1964). In acute hypoxia, at least up to 4300 m, $\dot{Q}$max does not differ substantially from sea-level controls (Saltin et al. 1968). On the basis of the described blood and hemodynamic adaptations, $\dot{V}_{O_2}$max is expected to decrease 30 to 40% in acute hypoxia as actually found. The expected drop in $\dot{V}_{O_2}$max should be less in chronic conditions, which is clearly not the case (see Fig. 4.2). Therefore, in recent years it has been hypothesized that in chronic hypoxia skeletal muscle mass and its function may be impaired.

4. Another experiment carried out originally at Mt. Everest base camp (Cerretelli 1976b) consisted of giving normoxic or slightly hyperoxic mixtures to breathe to 10 subjects fully acclimatized to high altitude and of measuring $\dot{V}_{O_2}$max in reference both to the sea-level prehypoxia values and to chronic hypoxia (Fig. 4.3). $\dot{V}_{O_2}$max determined at 5350 m

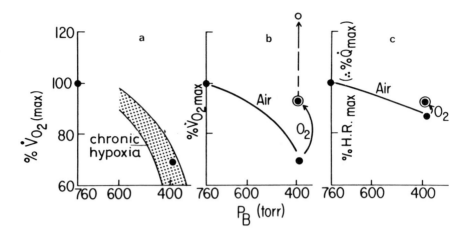

FIGURE 4.3. $\dot{V}O_2$max and HRmax (percentage of sea-level averages) as a function of PB. (a) The pointed area indicates the range of $\dot{V}O_2$max as a function of altitude (Fig. 2). The solid dot designates the average value found by Cerretelli (1976b) ($n = 10$). (b) The effect of breathing oxygen is indicated by the arrow. As an effect of normoxic or hyperoxic breathing, $\dot{V}O_2$max, which was reduced to 70% in absolute value (l·min⁻¹) by hypoxia, resumed only 92% of the sea-level control. The difference between control $\dot{V}O_2$max at sea level and $\dot{V}O_2$max at Mt. Everest base camp (PB ≅ 400 Torr) with supplemental oxygen was statistically significant (2P < 0.05, paired t-test); ○ indicates the expected $\dot{V}O_2$max value when breathing O_2 on the assumptions made in the text; (c) Effect of breathing oxygen on HRmax. (Modified from Cerretelli 1976b.)

by a closed-circuit method was found to be 2.26 l·min⁻¹, that is ~ 70% of the sea-level prehypoxia value (3.21 l·min⁻¹). When administering pure oxygen, it increased to 2.94 l·min⁻¹, that is only to 92% of the sea-level control value in spite of only a slight reduction of $\dot{Q}$max. The difference between the latter value and the sea-level reference data was statistically significant (2P<0.05, paired t-test). In particular, $\dot{Q}$ determined in two of the subjects appearing in Fig. 1 during an exercise requiring ~ 90% of $\dot{V}O_2$max at 5350 m was 19 and 20.7 l·min⁻¹, respectively, that is, 91 and 87% of the respective sea-level controls. The corresponding HR values were 148 and 162 beats·min⁻¹, respectively, that is, 83 and 91% of the values found at sea level. $\dot{Q}$max estimated by extrapolation of the individual $\dot{Q}$ vs. $\dot{V}O_2$ relationship to $\dot{V}O_2$max was found to be 10% less than in control conditions. Cardiac output was not measured during oxygen breathing. It appears conceivable, however, that $\dot{Q}$max$_{O_2}$ is, if anything, somewhat higher than $\dot{Q}$max$_{hyp}$. This is also compatible with the higher HRmax average levels attained while breathing O_2 (169 vs. 161 beats·min⁻¹) (see Fig. 4.3). Considering the 30–35% increase in [Hb] (Cerretelli 1976b) and assuming an average decrease in $\dot{Q}max$ of 10%, the maximal aerobic power of an acclimatized subject breathing O_2 at the base camp of Mt. Everest could approach a value 20–25% higher than at sea level. On the contrary, as previously pointed out and shown in Fig. 4.3, $\dot{V}o_2max$ attains only 92% of the sea-level control. This indicates that a defective step in the O_2 transport cascade does exist in the acclimatized individual. This reduction is localized most likely at the peripheral level.

The Effects of Chronic Hypoxia on Muscle Structure and Function

The main functional features and/or adaptational changes in hypoxia of the limb muscles of mountaineers were studied over the last decade in three groups of individuals:

One group (referred to as A) consisted of six elite high-altitude climbers, including Reinhold Messner and Peter Habeler, the first summiters on Everest without oxygen (Oelz et al. 1986). All these subjects had been repeatedly exposed to 8500 m or above without supplemental oxygen.
A second group (Group B) comprised the members of the 1981 Swiss Expedition to Mt. Lhotse Shar (8398 m) (Boutellier et al. 1983).
The third group (referred to as C) was made up of participants in the 1973 Italian expedition to Mt. Everest (Cerretelli et al. 1982).

The studies were conducted at sea level for Group A, and both at sea level (before departure and after return) and at altitude for Groups B and C.

Muscle Fiber Types and Morphometry of Elite Climbers

Muscle Fiber Types

The vastus lateralis muscle of elite climbers (Group A) contains a high percentage of type I fibers (Table 4.1). The mean value (70.2%) for this slow-twitch, fatigue-resistant fiber type is well beyond the average of 51% found with the same technique in sedentary men, but lower than the 78% observed in the best Swiss long-distance runners (Howald 1982). In two of the subjects the content of fast-twitch glycolytic type IIB fibers was much higher than expected for athletes of the endurance type.

Muscle Morphometry

The mean cross-sectional area of the thigh muscle fibers was significantly smaller in the group of outstanding climbers than in sedentary men of the

TABLE 4.1. Fiber type distribution in vastus lateralis muscle in elite high-altitude climbers.

Subjects	Type I (%)	Type IIA (%)	Type IIB (%)
RM	67.0	27.0	6.0
MD	72.0	28.0	0.0
DS	—	—	—
PH	70.0	17.0	13.0
HE	66.0	22.0	12.0
FM	76.0	18.0	6.0
Mean	70.2	22.4	7.4
SD	4.1	4.9	5.3

From Oelz et al. 1986.

same age, and it is only about 50% of the average fiber area measured in well-trained long-distance runners (Table 4.2). The mean diameter of the fibers is proportionally reduced. The number of capillaries per fiber is slightly greater than the average for sedentary subjects, but significantly smaller than for long-distance runners (1.7 : 2.7). However, since the muscle fibers of the climbers are much thinner (3100 : 6400 μm^2 of cross section) than those of untrained subjects or of endurance athletes, this is indicative of favorable conditions for tissue oxygenation.

The volume density of total and interfibrillar mitochondria in the Vastus lateralis muscle of the climbers is similar to that found by the same method in untrained men of the same age, but significantly lower than the

TABLE 4.2. Morphometry of capillaries and fibers in cross sections of Vastus lateralis muscle.

Subjects	$N_N(c,f)$ (unitless)	$N_A(c,f)$ (mm^{-2})	a(f) (μm^2)	D(f) (μm)	$A_N(f,c)$ (μm^2)
RM	1.28	365	3524	66.98	2740
MD	2.05	614	3332	65.13	1628
DS	—	—	—	—	—
PH	1.29	457	2829	60.01	2186
HE	1.77	606	2918	60.95	1651
FM	1.97	670	2935	61.13	1492
Mean	1.67	542	3108	62.84	1939
SD	0.37	127	303	3.04	520
Sedentary	1.39	387*	3640*	67.88	2300
Athletes	2.70*	431	6410*	90.30*	2320

$N_N(c,f)$ = number of capillaries per fiber. $N_A(c,f)$ = number of capillaries per area of tissue. a(f) = area of fibers. D(f) = mean fiber diameter. $A_N(f, c)$ = area of fibers perfused per capillary.
* Significantly different from climbers (unpaired t-test, 2P < 0.05).
From Oelz et al. 1986.

TABLE 4.3. Morphometry of the cellular ultrastructure in m. Vastus lateralis.

Subjects	Vv(mc,f) (%)	Vv(ms,f) (%)	Vv(mt,f) (%)	Vv(li,f) (%)	Vv(fi,f) (%)	Vv(re,f) (%)
RM	4.61	0.79	5.41	0.12	76.26	17.15
MD	4.36	1.11	5.46	0.59	78.78	15.16
DS	—	—	—	—	—	—
PH	3.89	0.64	4.53	1.73	78.28	15.47
HE	4.32	0.46	4.78	1.11	82.03	12.08
FM	3.85	0.71	4.56	0.61	81.39	13.45
Mean	4.21	0.74	4.95	0.83	79.35	14.66
SD	0.33	0.24	0.46	0.61	2.36	1.95
Sedentary	4.25	0.48	4.74	0.68	82.79*	11.57
Athletes	6.57*	0.74	7.32*	0.85	82.97*	8.86*

Vv(mc,f) = volume density of interfibrillar mitochondria.
Vv(ms,f) = volume density of subsarcolemmal mitochondria.
Vv(mt,f) = volume density of total mitochondria.
Vv(li,f) = volume density of intracellular lipid droplets.
Vv(fi,f) = volume density of myofibrils.
Vv(re,f) = volume density of residual sarcoplasmic components.
* Significantly different from climbers (unpaired t-test, $2P < 0.05$).
From Oelz et al. 1986.

values observed in well-trained long-distance runners (Table 4.3). As is well known, there is a direct relationship between these figures and the measurements of maximum oxygen uptake (Table 4.4), thus confirming the importance of the mitochondrial volume as one of the determinants of the maximum aerobic power in humans (Hoppeler et al. 1985). The

TABLE 4.4. Oxygen uptake, heart rates, and peak lactate concentrations in two different treadmill tests.

Subjects	$\dot{V}O_2$ r 10 (ml·min⁻¹·kg⁻¹)	HR r 10 (min⁻¹)	$\dot{V}O_2$ w 35 (ml·min⁻¹·kg⁻¹)	HR w 35 (min⁻¹)	$\dot{V}O_2$max (ml·min⁻¹·kg⁻¹)	[Lâ_b] (mM)
RM	48.8	184.0	46.6	161.0	48.8	15.3
MD	60.8	192.0	57.0	180.0	60.8	12.2
DS	57.0	181.0	63.0	174.0	63.0	15.1
PH	65.9	182.0	55.4	172.0	65.9	12.8
HE	54.5	195.0	56.1	185.0	56.1	7.2
FM	62.5	203.0	58.0	187.0	62.5	8.6
Mean	58.3	189.5	56.0	176.5	59.5	11.9
SD	6.1	8.7	5.3	9.6	6.2	3.3

——————NS——————

——————— $2P < 0.005$ ———————

$\dot{V}O_2$ r 10 = peak $\dot{V}O_2$ run 10% grade. HR r 10 = peak heart rate run 10% grade. $\dot{V}O_2$ w 35 = peak $\dot{V}O_2$ walk 35% grade. HR w 35 = peak heart rate walk 35% grade. [Lâ_b] = maximum blood lactate concentration at exhaustion.
2P by paired t-test.
From Oelz et al., 1986.

volume density of subsarcolemmal mitochondria and of intracellular lipid droplets in the muscles of the climbers is as high as for elite long-distance runners (Hoppeler et al. 1973, 1985). Since these variables probably reflect the muscle fiber's capacity to oxidize fat, the present results indicate that outstanding high-altitude climbers may rely heavily on fat as an energy source during exercise.

MORPHOLOGICAL AND BIOCHEMICAL ADJUSTMENTS OF LIMB MUSCLES AFTER PROLONGED EXPOSURE TO CHRONIC HYPOXIA

This problem was investigated on seven members of the 1981 Swiss Mt. Lhotse expedition (Group B). Biopsies were taken from the Vastus lateralis muscle immediately before departure and upon return from the expedition, within 2 weeks from leaving base camp (5200 m).

The measurements show (Table 4.5) that the ratio of the number of capillaries to the number of fibers was unchanged (2.25 vs. 2.30) throughout the hypoxic exposure. By contrast, the mean diameter of the fibers was slightly but significantly reduced from 78.1 to 71.5 μm, which implies a calculated concomitant decrease of muscle mass of about 10–15%. As indicated before, the decrease of muscle mass found by computer tomography of the thigh was on the order of 12%. Biochemically, a 34% decrease of muscle proteins (expressed as gram of protein per unit of muscle wet weight, not shown in the table) was also found. The described structural changes, on the one hand, are expected to impair the muscle mechanical performance, but on the other, they should improve O_2 extraction due to a shorter O_2 diffusion path. With regard to enzymatic changes (Table 4.5), the activity of succinate dehydrogenase (SDH), a key enzyme of the tricarboxylic acid (TCA) cycle, which reflects the adaptations of the respiratory status of the muscles to physiological and pathological stress, was found to be on the average 45% lower after return from altitude in three subjects. Similarly, the activity of phosphofructokinase (PFK) and of lactate dehydrogenase (LDH) was found to be significantly reduced, which may have some bearing on muscle lactacid capacity and power. Whether the described changes are the results of reduced muscle activity due to hypoxia or of hypoxia per se is still a matter for investigation. Recent measurements carried out in a group of 7 members of the 1986 Swiss Mt Everest expedition have confirmed the decrease of muscle mass and of the mitochondrial-to-fiber volume ratio ($\sim$ 25%) 2 to 3 weeks after leaving altitude. The activity of some enzymes of the TCA cycle (citrate synthase, malate dehydrogenase) was also $\sim$ 20% reduced. At variance with the data of Table 4.5, PFK and LDH were unchanged in this group of particularly well trained individuals (H. Howald, personal communication.)

TABLE 4.5. Muscle morphological and functional changes upon exposure to high altitude.

	NC/Nf ($n°$)	Øf (μm)	SDH	PFK	LDH
			mM·min⁻¹·kg⁻¹muscle		
B	2.25 ± 0.22	78.1 ± 4.4	7.0 ± 2.2	70 ± 14	460 ± 118
A	2.30 ± 0.25	71.5 ± 6.2	4.5 ± 0.5	50 ± 20	230 ± 128
Significance of the difference	NS	2P<0.05	2P<0.2	2P<0.1	2P<0.005
n (subjects)	7	7	3	7	7

B = before; A = 12 days after 5 weeks at 5200 m or above. Nc = number of capillaries; Nf = number of fibers; Øf = diameter of fibers; NS = not significant.

AFTEREFFECTS OF CHRONIC HYPOXIA ON CARDIAC OUTPUT AND MUSCLE MICROCIRCULATION

The Adjustment Rate of Muscle Blood Flow ($\dot{Q}m$)

The kinetics of the $\dot{Q}m$ readjustment upon square-wave exercises of various loads measured from the time (sec) required for blood flow to attain steady state has been studied on several members of Group B, and the results are shown in Table 4.6 (Cerretelli et al. 1984). On the whole, it may be seen that after hypoxia there is a tendency toward a slower response, particularly at the lower work loads and for the arm muscles

TABLE 4.6. Time (sec) required for $\dot{Q}$m to attain steady state following a square-wave increase of $\dot{V}$ power (W) before and after exposure to hypoxia.

| | Biceps | | Vastus lateralis | |
	Rest → 50 W		Rest → 75 W	Rest → 125 W
		(s)		
Control	16.5 ± 16.8*		17.1 ± 9.9†	31.5 ± 14.4
After hypoxia	56.2 ± 29.3*		28.8 ± 9.3†	33.1 ± 14.6
Measurements	6		14	12
Subjects	3		8	6

* $P < 0.05$.
† $P < 0.025$.
From Cerretelli et al. 1984.

(biceps). Even so, the kinetics of blood flow readjustment is 2 to 4 times faster than that of the $\dot{V}o_2$on response (see Fig.4.4) and, therefore, cannot be held responsible for the increase of the O_2 deficit found upon onset of exercise.

Cardiac Output ($\dot{Q}$) and Muscle Blood Flow ($\dot{Q}$m)

Upon return to sea level, the cardiac output ($\dot{Q}$) of the members of Group B was unchanged at rest as well as during cycloergometric exercise at 75 W, but was about 20% less than in the controls at 150 W (19.6 vs. 24.9 $1 \cdot min^{-1}$) in spite of a nearly constant HR (approx. 137 beats·min^{-1}). $\dot{Q}$m in the vastus lateralis muscle at rest was identical with the control, but was considerably reduced both at 75 W (21.9 vs. 29.6 ml·100 $g^{-1} \cdot min^{-1}$) and at 125 W (28.4 vs. 46.7). The drop of $\dot{Q}$m exceeded greatly the concomitant decrease of $\dot{Q}$. Thus, the reduction of nutritional blood flow to the muscles appears to be the result mainly of a local change (not at the capillary level), perhaps of the opening of arteriovenous anastomoses. Such hypothetical adaptation implies that a fraction of muscle blood flow bypasses the metabolically acive districts, thus reducing the load on the heart but also curtailing the maximal amount of O_2 available to the contracting muscles. Despite this, submaximal work can still be sustained thanks to the shorter diffusion distance for O_2 described above. The present results on $\dot{Q}$m are in agreement with the data of Bidart et al. (1975) showing that altitude natives as well as acclimatized lowlanders are characterized by a lower muscle blood flow per unit muscle mass for a given work load. These observations support the hypothesis that reduced muscle perfusion may be one of the main factors affecting $\dot{V}o_2$max at altitude.

The Measurement of Metabolic Transients before and after Exposure to High Altitude. The $\dot{V}_{O_2}$ON and $\dot{V}_{O_2}$OFF Responses

The $\dot{V}_{O_2}$on response as measured at the mouth at the beginning of muscular exercise lags behind the performance of mechanical work. As a consequence, at the onset of exercise, the O_2 uptake through the mouth is insufficient to meet the energy requirement of the working muscles (an O_2 "deficit" is incurred), while at the end it is in excess (an O_2 "debt" is paid). During the rest-to-work transient, three different processes can provide the missing fraction of energy required by the working muscles that is not available from external gas exchange: (1) phosphocreatine (PC) breakdown, (2) anaerobic glycolysis with lactic acid (La) formation ("early lactate," Cerretelli et al. 1979), and (3) depletion of O_2 stores. The O_2 deficit can therefore be viewed as the sum of at least three terms (di Prampero et al. 1983):

$$O_2 \text{ deficit} = V_{O_2}^{PC} + V_{O_2}^{eLa} + \Delta V_{O_2}st \qquad (4.1)$$

where $V_{O_2}^{PC}$ = the O_2 equivalent of PC breakdown

$V_{O_2}^{eLa}$ = the O_2 equivalent of the net amount of La produced before the attainment of the steady state

$\Delta V_{O_2}st$ = the O_2 taken from body stores (blood, myoglobin, alveolar air)

In the recovery phase after exercise, the PC concentration and the body O_2 stores are rapidly rebuilt to preexercise levels. On the contrary, the removal of La proceeds at a much slower rate, La being either oxidized as substrate or reconverted to glycogen. As a consequence, the O_2 debt paid is given by the sum of only two terms:

$$O_2 \text{ debt} = V_{O_2}^{PC} + \Delta V_{O_2}st \qquad (4.2)$$

As shown above, chronic hypoxia induces important changes in muscle structure and function (see Table 4.5) which are likely to affect the different components of the O_2 deficit and debt as defined in Eqs. (1.1) and 1.2). Therefore single-breath $\dot{V}_{O_2}$ on and off curves were determined at rest and during 125-W cycloergometric efforts on seven members of Group B before departure and 12–14 days after return from the Mt. Lhotse Shar expedition. The subjects were exposed for 8 weeks between 3500 and 8398 m.

The data for a typical subject are shown in Fig. 4.4A and B. It appears that the shape of the $\dot{V}_{O_2}$ on response curve is a sigmoid (from monoexponential), while the time required to attain 50% of the $\dot{V}_{O_2}$on response increases (40.8 vs. 30.5 sec, $P < 0.01$, $n = 7$) (Boutellier et al. 1984). Among the factors possibly responsible for this finding, $V_{O_2}^{PC}$ and $V_{O_2}^{eLa}$ can be discarded since the muscle PC stores were probably unchanged

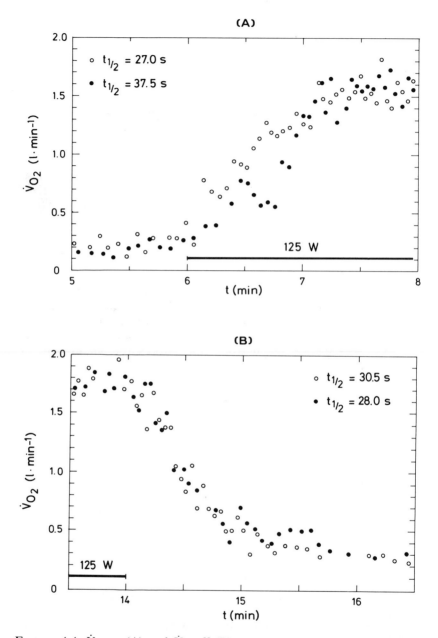

FIGURE 4.4. $\dot{V}_{O_2}$on (A) and $\dot{V}_{O_2}$off (B) response at the onset and offset of a rectangular 125-W cycloergometric exercise before (○) and after (●) prolonged exposure to hypoxia. (From Boutellier et al. 1984.)

after hypoxia (see below) and La production was not significantly different in the two investigated conditions; by contrast, the term $\Delta V_{O_2}st$ is most likely responsible for the observed change that is attributable to the modification of the hemoglobin- and/or myoglobin-bound O_2 stores. Another possible origin for the observed change of kinetics of the $\dot{V}_{O_2}$on response curve could be the biochemical adaptations occurring in muscle following hypoxia (see below and Table 4.5).

With regard to the O_2 debt payment, the $\dot{V}_{O_2}$ off curves shown in Fig. 4.4 do not differ significantly within the first 5 min of recovery.

MUSCLE ANAEROBIC METABOLISM AT ALTITUDE

The Glycolytic Contribution during a Supramaximal Exercise at 5200 m

Peak blood lactate concentration above rest $[\Delta L\hat{a}_b]$ was determined in four subjects (members of Group B) before and after 3 weeks at 5200 m or above, following 5, 10, 15, and 20 maximal standing jumps off both feet at the fastest possible rate on a force platform. As appears from Figs. 4.5 (di Prampero et al., personal communication 1985), La accumulation as a function of time (and/or of the cumulative number of jumps) is less at altitude than at sea level, indicating an impairment in the mobilization of the anaerobic glycolytic energy sources. This is possibly related to the reduced PFK and LDH activity in the muscle (see Table 4.5). Moreover, the onset of anaerobic glycolysis is delayed from about 3 to 12 sec. At exhaustion, the alactic energy stores account in hypoxia for at least 350 J/kg of body weight in equivalent oxidative energy, i.e., for the same amount observed at sea level. These observations indicate that the maximal absolute peak power of the subjects should not be affected by altitude (see below).

The Maximal Lactacid Capacity at Altitude

The effects of altitude on anaerobic glycolysis and the building of lactic acid in blood are fairly well documented by the recent studies of Cerretelli et al. (1982) on Group C.

Resting blood lactate concentration $[La_b]$ is practically unchanged throughout acclimatization to hypoxia. By contrast, the peak La_b concentration after exhausting exercise $[L\hat{a}_b]$, as shown by Cerretelli et al. (1982), is severely reduced at altitude, particularly above 4000 m (Fig. 4.6). The line shown in Fig. 4.6 can be extended up to altitudes of 6500 m and above (West, personal communication 1985). This limitation may be due to several factors, the most probable of which are: (a) the increased drop of intracellular pH for a given La accumulation (reduced buffer power due to acclimatization) followed by an impairment of the function of one or more rate-limiting enzymes along the glycolytic sequence, most

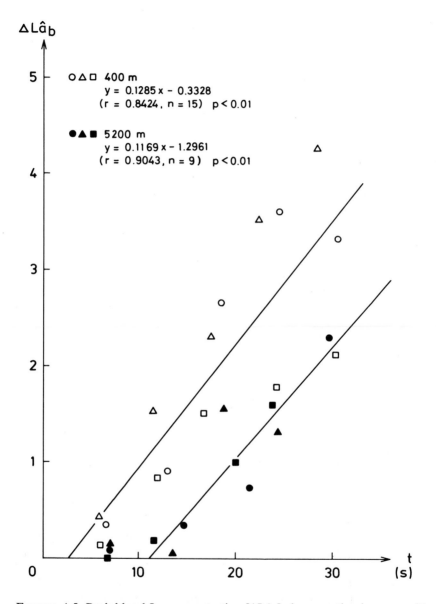

FIGURE 4.5. Peak blood La concentration [$\Delta L\hat{a}_b$] above resting in venous blood as a function of time (sec) following the performance of 5, 10, 15, and 20 maximal jumps (see text). Closed symbols: After 3 weeks at 5200 m and above. Open symbols: Control at sea level ($n = 4$).

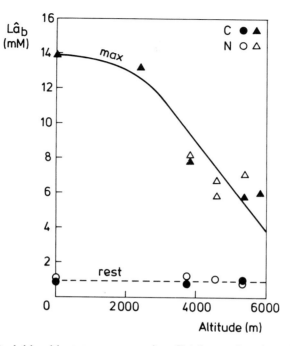

FIGURE 4.6. Peak blood lactate concentration, [Lâ$_b$], as a function of altitude. ---
indicates resting values; ___ indicates maximal exercise values. ●, ▲ indicate
Caucasians; ○, △ indicate high-altitude natives (Sherpas and Peruvian Indians).
(Modified after Cerretelli et al. 1982.)

likely phosphofructokinase (PFK); and (*b*) the reduced availability of
muscle glycogen.

The Alactic Mechanism: Effects of Chronic Hypoxia on Maximal Muscular Power ($\hat{w}_{Al}max$)

To the authors' knowledge, only two investigations have been devoted to
the study of the effects of altitude on maximal anaerobic power. In the
first one (di Prampero et al. 1982), the effects of acute and chronic
hypoxia on the "average" power output (i.e., the maximal power that can
be sustained for 3–5 sec, $\bar{w}_{Al}max$) were investigated. In the second study,
which was carried out on members of Group B during the 1981 Swiss
Lhotse Shar expedition, the effects of high-altitude acclimatization on
"instantaneous" power ($\hat{w}_{Al}max$) were assessed (unpublished data). The
results of these studies are reported in Table 4.7. It appears that (1) acute
hypoxia (14.5% O_2 in N_2) has no substantial effects on either $\bar{w}_{Al}max$ or
$\hat{w}_{Al}max$; (2) chronic hypoxia (3 weeks at P$_B$ of 405–430 mmHg) does not
produce any detectable change of either $\bar{w}_{Al}max$ or $\hat{w}_{Al}max$; and (3)

TABLE 4.7. Maximal mechanical power during a 10-sec all-out cycloergometric effort ($\bar{w}_{Al}$max, W·kg⁻¹) and during the pushing phase of a standing high jump off both feet ($\hat{w}$ average of first 5 jumps in a row).

PB (mmHg)	750	750	750	430
Altitude (m)	Sea level	Sea level	Sea level	4.540
$F_{I_{O_2}}$	1.00	0.145	0.121	0.209
Time at altitude	—	—	—	3 weeks
Load (N)	53.6±3.5	54.5±2.0	52.8±2.0	52.9±2.0
$\bar{w}_{Al}$max(W·kg⁻¹)	11.1±1.2	11.3±1.0	11.2±1.00	11.0±0.9
n	24	24	24	24
PB (mmHg)	730	405	405	730
Altitude (m)	400	5.200	5.200	400
Time at altitude	Before departure	3 weeks	5 weeks	Immediately after return
$\hat{w}_{Al}$max(W·kg⁻¹)	21.0±1.9	20.8±3.9	16.0±1.8	17.4±1.5
n	15	25	15	20

$\hat{w}_{Al}$max (W·kg⁻¹) is reported for several subjects in different experimental conditions. The load (N) at which the maximal power during the cycloergometric effort was attained is also indicated. Average = ±1 SD values; n = number of measurements.
From di Prampero et al. 1982.

chronic hypoxia (>5 weeks at PB = 405 mmHg and below) reduces significantly $\hat{w}_{Al}$max.

As is well known, at the beginning of muscular exercise, O_2 consumption and La production do not contribute significantly to the energy requirement of the working muscles since their time course is rather slow compared with the mechanical events of the contraction. As a consequence, both "average" and "instantaneous" methods for assessing maximal muscular power output provide a measure proportional to the maximal rate of $\sim$ P splitting in muscle (di Prampero 1981).

It can be concluded that neither acute nor chronic hypoxia up to 3 weeks' duration has any effects on the maximal rate of $\sim$ P splitting in muscle, at least up to 5200 m above sea level. This finding is in agreement with the data of Knuttgen and Saltin (1973), who have shown that ATP and PC concentrations are not affected by acute hypoxia, at least up to a simulated altitude of 4000 m above sea level. Raynaud and Durand (1982) have also shown that the O_2 debt paid within the first minute of recovery after 3 weeks at 3800 m above sea level is essentially the same as at sea level. Since the O_2 debt paid during the first minute can be taken as representative of $\sim$ P resynthesis after exercise, these data suggest that hypoxia has no major effects on this aspect of muscle metabolism. From the data of Table 4.7 it appears, however, that for altitudes > 5200 m and for prolonged exposures to hypoxia, $\hat{w}_{Al}$max may undergo a substantial drop, probably a consequence of muscle deterioration (Table 4.5).

Conclusions

The growing experimental evidence that both aerobic and anaerobic energetic mechanisms, particularly those of the skeletal muscle, are affected negatively by prolonged exposure to high altitude is rather impressive. The morphological and functional changes of the muscle described can explain satisfactorily the cause of the deterioration of exercise performance of acclimatized individuals.

Unfortunately, the role of most of the variables underlying such changes is still unknown. Among these, the degree, duration, and mode of exposure; the species specificity; and the effect of genetic variables, of training, of the decrease of body buffers, and, particularly, of malnutrition are only the most common ones. What for several decades has been attributed exclusively to lack of availability of substrates and converter (O_2) must now be reconsidered in the light of more profound alterations of some essential biochemical pathways. This more comprehensive approach may also prove suitable to shed some light into the pathophysiology of several hypoxic diseases.

REFERENCES

Bidart Y, Druet L, Durand J (1975). Débit dans le muscle squelettique chez les sujets résidants et transplantés en altitude (3800m). *J Physiol (Paris)* 70: 333–337

Boutellier U, Giezendanner D, Cerretelli P, di Prampero PE (1984). After effects of chronic hypoxia on $\dot{V}O_2$ kinetics and on O_2 deficit and debt. *Eur J Appl Physiol* 53: 87–91

Boutellier U, Howald H, di Prampero PE, Giezendanner D, Cerretelli P (1983). Human muscle adaptations to chronic hypoxia. In: Sutton JR, Houston CS, Jones NL (eds) *Hypoxia, Exercise and Altitude.* AR Liss, New York, pp 273–281

Cerretelli P (1976a). Metabolismo ossidativo e anaerobico nel soggetto acclimatato all'altitudine. *Minerva Aerosp* 67: 11–26

Cerretelli P (1976b). Limiting factors to oxygen transport on Mount Everest. *J Appl Physiol* 40: 658–667

Cerretelli P, di Prampero PE (1985). Aerobic and anaerobic metabolism during exercise at altitude. In: Jokl E, Hebbelinck M. (eds) *Medicine Sport Science.* Karger, Basel, vol 19, 1–19

Cerretelli P, Marconi C, Dériaz O, Giezendanner D (1984). After effects of chronic hypoxia on cardiac output and muscle blood flow at rest and exercise. *Eur J Appl Physiol* 53: 92–96

Cerretelli P, Pendergast D, Paganelli WC, Rennie DW (1979). Effects of specific muscle training on $\dot{V}O_2$on- response and early blood lactate. *J Appl Physiol* 47: 761–769

Cerretelli P, Veicsteinas A, Marconi C (1982). Anaerobic Metabolism at High Altitude: The Lactacid Mechanism in: Brendel W, Zink RA (eds) *High altitude Physiology and Medicine.* Springer, New York, pp 95–102 di Prampero PE (1981). Energetics of muscular exercise. *Rev Physiol Biochem Pharmacol* 89: 143–222

di Prampero PE, Boutellier U, Pietsch P (1983). O_2 deficit and stores at onset of muscular exercise in humans. *J Appl Physiol* 55: 146–153

di Prampero PE, Mognoni P, Veicsteinas A (1982). The effects of hypoxia on maximal anaerobic alactic power in man. In: Brendel W, Zink RA (eds) *High Altitude Physiology and Medicine*. Springer, New York, pp 88–93

Hoppeler H, Howald H, Conley K, Linstedt L, Claassen H, Vock P, Weibel ER (1985). Endurance training in humans: Aerobic capacity and structure of skeletal muscle. *J Appl Physiol* 59: 320–327

Hoppeler H, Lüthi P, Claassen H, Weibel ER, Howald H (1973). The ultrastructure of the normal human skeletal muscle; a morphometric analysis of untrained men, women, and well trained orienteers. *Pflügers Arch* 344: 217–232

Howald H (1982). Training-induced morphological and functional changes in skeletal muscle. *Int J Sports Med* 3: 1–12

Knuttgen HG, Saltin B (1973). Oxygen uptake, muscle high energy phosphates, and lactate in exercise under acute hypoxic conditions in man. *Acta Physiol Scand* 87: 368–376

Oelz O, Howald H, di Prampero PE, Hoppeler H, Claassen H, Jenni R, Bühlmann A, Ferretti G, Brückner JC, Veicsteinas A, Gussoni M, Cerretelli P (1986). Physiological profile of world class high altitude climbers. *J Appl Physiol* 60: 1734–1742

Pugh LGCE (1964). Cardiac output in muscular exercise at 5800 m (19,000 ft) *J Appl Physiol* 19: 441–447

Raynaud, J, Durand J (1982). Oxygen deficit and debt in submaximal exercise at sea level and high altitude. In: Brendel W, Zink RA (eds) *High Altitude Physiology and Medicine*. Springer, New York, pp 103–106

Saltin B, Grover RF, Blomqvist CG, Hartley LH, Johnson RL Jr (1968). Maximal oxygen uptake and cardiac output after two weeks at 4300 m. *J Appl Physiol* 25: 400–409

Part II   Physiology of Diving and Exposure to Elevated Pressure

5

Resistance and Inertance When Breathing Dense Gas

Hugh D. Van Liew

Introduction

Inertance of the human respiratory system is negligible in normal environments, but has been shown experimentally to increase in simple direct proportion to the density of the breathing gas (Mead 1956; Peterson and Wright 1976; Sharp et al. 1964). This raises the question of how inertance in dense-gas environments compares with airflow resistance, which is also density-dependent. The pressure necessary to accelerate gas is greatest when flow is changing direction at end-inspiration and end-expiration, whereas an effect on pressure for gas flow is greatest during inspiration and expiration.

The pressure/flow relationship is nonlinear; Rohrer (1915) assumed that pressure for flow (P_{flow}) is equal to flow [the first derivative of volume (V) with respect to time (t)] multiplied by a constant ($K1$) plus (or minus) the square of flow multiplied by a second constant ($K2$)—the sign of the second term must be forced to be minus when flow is negative:

$$P_{flow} = K1 \, dV/dt + K2 \, (dV/dt)^2 \qquad (5.1)$$

Normal Density

Fig. 5.1 shows simulations of l-l breaths at the high frequency of 60 breaths/min in a normal-density environment. Calculations were done with an Apple MacIntosh microcomputer with a program in Microsoft Basic. Assumptions were (a) the equation of motion of the respiratory system (Mead 1961) for ventilating a simple, one-compartment human lung, (b) Eq. (1) for pressure to cause flow, and (c) arbitrary decisions or literature values for constants: total lung capacity—6 l; compliance of the total respiratory system—0.1 l/cm H_2O; inertance—0.01 cm $H_2O/l/sec^2$; $K1$—1.2 cm $H_2O/l/sec$; $K2$—0.28 cm $H_2O/(l/sec)^2$.

Volume was assumed to be a sine-wave function of time; this assignment makes flow a cosine function and acceleration a sine function. In the next-to-top panel of Fig. 5.1, pressure to overcome elastic recoil of the total respiratory system is seen to be in phase with volume, pressure for

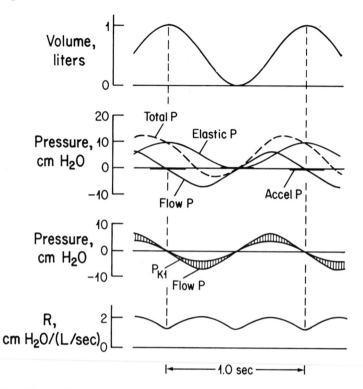

FIGURE 5.1. *Upper:* Lung volume above FRC vs. time; one-liter tidal volumes at 60/min in a normal environment. *Second:* Components of pressure; total pressure (dashed) at any time is the sum of pressures for overcoming elastic recoil, for flow, and for acceleration and deceleration of gas. Note that the convention in all figures is to plot as if pressure were being applied by a pump ventilator, so a positive pressure is an inflating pressure. *Third:* Components of pressure for flow; vertical distance from baseline to the curve labeled P_{K1} is pressure accounted for by the first term of Eq. (1); additional vertical distance (shaded) is pressure accounted for by the second term of Eq. (1). *Lowest:* Ratio of pressure for flow divided by flow.

flow is in phase with change of volume, and pressure for acceleration is 90° out of phase with flow, and therefore 180° out of phase with volume and elastic recoil pressure. Total pressure reaches its peak between the peaks for its two significant components and becomes negative toward the end of expiration—elastic recoil pressure assists expiration but is not great enough to achieve the mandated expiratory flow. For this particular frequency/tidal-volume combination, magnitude of accelerative pressure is so small that it hardly shows on the diagram.

The next-to-lowest panel in Fig. 5.1 separates the pressure accounted for by P_{K1}, the first term on the right side of Eq. 1.1, from the contribution of P_{K2}, the second term; for the mandated flows, peak P_{K1} is about the same magnitude as peak P_{K2}. The P_{K1} is a cosine wave, but the contribution of P_{K2} has a sharper configuration than a cosine wave because P_{K2} increases as the square of flow, making the total pressure for flow sharp also. The lowest panel in the figure shows the "resistance" calculated as pressure for flow divided by flow; the ratio is not constant because of the second term of Eq. (5.1).

High Density

Fig. 5.2 shows the pressure components for the same breaths as in Fig. 5.1, but with density increased, using the assumption that $K2$ of the Rohrer equation increases in proportion to density (Wood et al. 1976). Elastic pressure is the same as in Fig. 5.1 because tidal volume is the same, but peak-to-peak total pressures are now 60 and 100 cm H_2O. Pressure for acceleration (shaded for emphasis) is of consequence in these high-density environments; it approximates the P_{K1} component. However, the main reason for the high total pressures is the enlarged flow pressure. The P_{K1} component is the same as in Fig. 5.1, but the shaded P_{K2} contribution is increased 10- and 20-fold because of the density dependence of P_{K2}.

For the 20-times density case (Fig. 5.2, lower), the amplitudes of the accelerative pressure and elastic pressure are about the same. Since the two are out of phase with each other, they tend to cancel; the frequency at which they exactly cancel is known as the natural frequency or resonant frequency. At normal densities, the natural frequency of the human respiratory system is at 5 or 6 breaths/sec (DuBois et al. 1956); it decreases when density increases and is near 1 breath/sec at the density shown (Fig. 5.2, lower). At the natural frequency, pressure for flow is the only energy-dissipating entity, so total pressure is in phase with flow, as seen in the figure. Note, however, that pressure for flow is so large that it would have dominated the total pressure even if accelerative pressure were negligible; total pressure tends to be in phase with flow whenever flow pressure is large—the contribution of high accelerative pressure was to accentuate this tendency slightly (Van Liew 1987a).

Figs. 5.1 and 5.2 portray large, high-frequency breaths to emphasize the roles of accelerative and P_{K2} pressures. In low-frequency breathing, the low flows diminish the role of P_{K2} pressure so total pressure is dominated by elastic recoil with a flow contribution that is due mainly to P_{K1}. In low-tidal-volume/high-frequency breathing, P_{K2} would again be unimportant if tidal volumes are so small that flows are small, but accelerative pressure could be appreciable.

Hesser et al. (1981) found averages of tidal volume and frequency to be 2.4 l and 33/min when men performed maximal exercise while breathing air at 6 ATA. Fig. 5.3 shows simulations, using the same assumptions as

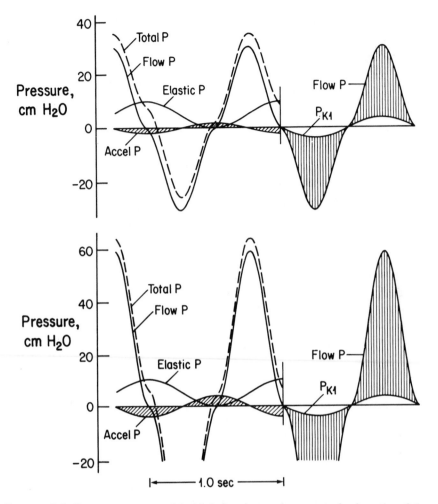

FIGURE 5.2. Pressures expected in high-density environments for breaths of the same tidal volume and frequency as in Fig. 5.1. Gas density is increased by factors of 10 (*upper*) or 20 (*lower*), as if men were breathing air at 10 and 20 ATA. The two components of Eq. (1) are shown alone at the right of the graphs. Pressures went off scale at the bottom.

in Figs. 5.1 and 5.2, for this real situation. Elastic pressure is larger than in Figs. 5.1 and 5.2 because tidal volume is larger; here it is about the same magnitude as flow pressure. Accelerative pressure is negligible in comparison. The total pressures (50 cm H_2O on inspiration, 25 cm H_2O on expiration) compare well with the intrapleural pressures measured in the study.

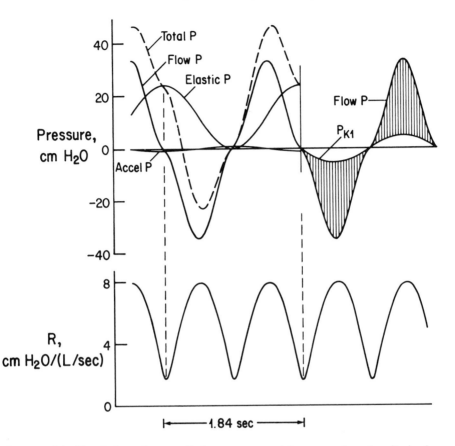

FIGURE 5.3. Simulations for a realistic case—exercising men when density is six times greater than normal. *Upper:* Components of total pressure, with components of flow pressure at the right. *Lower:* Ratio of flow pressure to flow.

Minute ventilation for this case is 79 l/min, close to the 60 l/min of the arbitrary patterns chosen for Figs. 5.1 and 5.2; because flows are about the same, P_{K1} patterns are similar also. The density dependence of P_{K2} makes it the largest component of pressure (Fig. 5.3, right side). Simulations for the same tidal volume and frequency at normal gas density showed pressure for flow to be much less; in normal density, total pressure for this breathing pattern is dominated by elastic recoil pressure and is nearly in phase with it.

The ratio of flow pressure to flow fluctuates very markedly when P_{K2} is an important component (Fig. 5.3, lower panel). Clearly, any idea of a certain unchanging airway resistance must be abandoned if the ratio changes from 2 to 8, as in the figure.

Critique of Assumptions

The assumption that all density dependence for flow resides in the $K2$ term of the Rohrer equation is debatable (Clarke et al. 1982; Wood et al. 1976). Pedley et al. (1970) gave theoretical reasons why pressure for flow (both $K1$ and $K2$) should increase as the square root of density, an idea that can be supported by experimental data (Wood and Bryan 1978). Of course, in simulations based on the square root assumption or other assumptions, major changes occur in the partitioning of flow pressure into its $K1$ and $K2$ components. However, when alternative assumptions are based on realistic coefficients that have been derived from experimental data, they make only minor changes in the patterns and magnitudes of total pressures and pressure-to-flow ratios for simulations resembling those shown in Figures 5.1 to 5.3 (Van Liew, 1987b). The assumption that all density dependence resides in the $K2$ term may have an advantage for didactic purposes in that density and flow rate effects are manifested in the one P_{K2} term.

I purposely left out issues that would have made the simulations more realistic, but also more complex than needed to illustrate the points under consideration. (a) Omission of the nonlinearity of the pressure-volume characteristics of the respiratory system causes peak elastic pressures to be underestimated, especially for large tidal volumes. (b) The tendency for chokes to develop in lung airways increases with density (Van Liew 1983). Flow through choked airways is approximately inversely proportional to the square root of gas density; omission of the phenomenon of dynamic compression means that volume changes brought about by given pressures during expiration are unrealistically high in the figures. (c) Pressure for flow is dependent on lung volume (Bouhuys and Jonson 1967); the magnitude of the effect can be shown to be of the order of 10 or 20% of the pressures shown on the figures. (d) Inertia of body structures, viscosity changes with density, tissue resistance, laryngeal or nasal resistance, and deviations of actual volume-vs.-time patterns from sine waves were all assumed to be negligible.

Conclusions

These extrapolations to dense-gas environments, as well as other simulations not reported here, make it appear that pressure for acceleration is never an important fraction of total pressure for breathing. An increase of accelerative pressure has a minor effect on phase relations between pressure and flow in the same direction that high flow resistance does—to make intrapleural pressure swings tend to be in phase with flow. The simulations do not rule out the possibility of second-order effects of accelerative phenomena, such as enhancement of ventilation to lung regions that are served by straight paths in the airways (Clarke et al. 1981) or production of relatively high peak pressures in the alveolar regions (Fredberg et al. 1984).

In general, high frequency and high density, the two variables that increase accelerative pressure, also increase flow pressure so much that it dominates the energy requirement for breathing during exercise in dense-gas environments.

REFERENCES

Bouhuys A, Jonson B (1967). Alveolar pressure, airflow rate, and lung inflation in man. *J Appl Physiol* 22: 1086–1100

Clarke JR, Fisher MA, Jaeger MJ (1981). Inertance as a factor in uneven ventilation in diving. In: Bachrach AJ, Matzen MM (eds) *Underwater Physiology VII*. Undersea Medical Society, Bethesda, MD, pp 225–233

Clarke JR, Jaeger MJ, Zumrick JL, O'Bryan R, Spaur WH (1982). Respiratory resistance from 1 to 46 ATA measured with the interrupter technique. *J Appl Physiol: Respirat Environ Exercise Physiol* 52: 549–555

DuBois AB, Brody AW, Lewis DH, Burgess, BF Jr (1956). Oscillation mechanics of lungs and chest in man. *J Appl Physiol* 8: 587–594

Fredberg JJ, Keefe DH, Glass GM, Castile RG, Franz ID III (1984). Alveolar pressure nonhomogeneity during small-amplitude high-frequency oscillation. *J Appl Physiol: Respirat Environ Exercise Physiol* 57: 788–800

Hesser CM, Linnarsson D, Fagraeus L (1981). Pulmonary mechanics and work of breathing at maximal ventilation and raised air pressure. *J Appl Physiol: Respirat Environ Exercise Physiol* 50: 747–753

Mead J (1956). Measurement of inertia of the lungs at increased ambient pressures. *J Appl Physiol* 9: 208–212

Mead J (1961). Mechanical properties of lungs. *Physiol Rev* 41: 281–330

Pedley TJ, Schroter RC, Sudlow MF (1970). The prediction of pressure drop and variation of resistance within the human bronchial airways. *Respir Physiol* 9: 387–405

Peterson RE, Wright WB (1976). Pulmonary mechanical functions in man breathing dense gas mixtures at high ambient pressures—*Predictive Studies II*. In: Lambertsen CJ (ed) *Underwater Physiology V*. FASEB, Bethesda, MD, pp 67–77

Rohrer, F (1915). Der Stroemungswiderstand in den menschlichen Atemwegen. *Pflügers Arch Physiol* 162: 225–259

Sharp JT, Henry JP, Sweany SK, Meadows WR, Pietras RJ (1964). Total respiratory inertance and its gas and tissue components in normal and obese men. *J Clin Invest* 43: 503–509

Van Liew HD (1983). Mechanical and physical factors in lung function during work in dense environments. *Undersea Biomed Res* 10: 255–264

Van Liew HD (1987a) The electrical-respiratory analogy when gas density is high. *Undersea Biomed* Res 14: 149–160

Van Liew HD (1987b) Components of the pressure required to breathe dense gases. *Undersea Biomed Res* 14: 263–276

Wood LDH, Bryan AC (1978). Exercise ventilatory mechanics at increased ambient pressure. *J Appl Physiol: Respirat Environ Exercise Physiol* 44: 231–237

Wood LDH, Engel LA, Griffin P, Despas P, Macklem PT (1976). Effect of gas physical properties and flow on lower pulmonary resistance. *J Appl Physiol* 41: 234–244

6

Diffusive Gas Mixing in the Lungs: A Possible Factor Limiting Alveolar Gas Exchange at Depth

Y. Ohta and L.E. Farhi

There is conclusive evidence that diffusion in the gas phase may play an important role in gas transport (Cumming et al. 1967; Georg et al. 1965; Okubo and Piiper 1974; Worth et al. 1977). In this chapter we will examine the role of diffusion in alveolar gas exchange on the basis of (1) work with a physical model of stratification in the gas phase and (2) data on gas exchange across the alveolar-capillary membrane obtained from animal and human experimentation. Using that information, we will conclude with an analysis of hypoxemia occurring while breathing a normoxic mixture in a hyperbaric environment, a phenomenon first described by Chouteau (1969) and Chouteau et al. (1971). Our thesis is that since diffusion is influenced by the molecular composition of the gas mixture and by ambient pressure, it may become a limiting factor in alveolar gas exchange at depth.

A Model Experiment

We experimented with a model designed to demonstrate the difference in the magnitude of stratification of gases of different diffusivities.

Fig. 6.1 shows the experimental system used in the present study. Two large acrylic cylinders make a water-sealed gas phase of approximately 1900 ml, in which a hollow bronchial cast of a mongrel dog is hung on a straight tube with a cross-sectional area of 0.95 cm^2. The speed and extent of movement of the bottom cylinder and the pause time at end-expiration can be changed so as to simulate a variety of ventilatory patterns.

The hollow bronchial cast used in the present study stops at the sixth generation and is trimmed so that each branch has approximately equal length. Two electric fans inside the chamber can be used to stir gas and thereby prevent stratification in the static gas phase. Helium, argon, and sulfur hexafluoride were used as indicator gases and were measured by a mass spectrometer (Varian, MAT M-3, West Germany).

We conducted two kinds of experiments. In one type of run, the mass

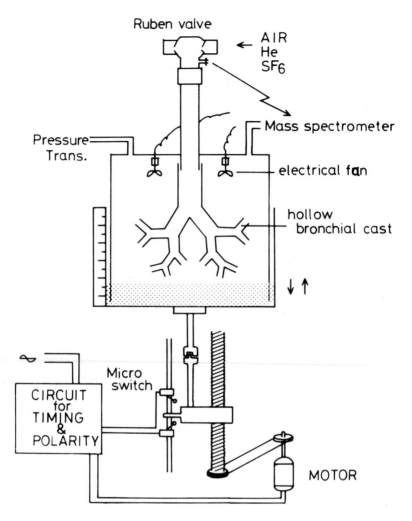

FIGURE 6.1. Schematic diagram of the experimental model system

spectrometer sampling tube was fixed near the top of the box so that the distance between the sampling site and the distal tip of the bronchial cast was fixed at approximately 10 cm. Gas fractions at that single point are studied as a function of time. In the other runs, the gas samples were taken at the various points between the distal end of the cast and the top of the box, and the measurements are reported as a function of the distance.

STRATIFICATION AS A FUNCTION OF TIME

Fig. 6.2 is a typical experimental record of simultaneous washin of 2% helium and SF_6 in nitrogen, respectively, into nitrogen in the box. The

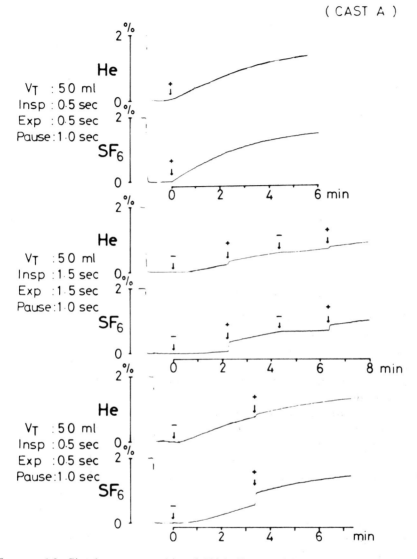

FIGURE 6.2. Simultaneous washin of 2% helium and SF$_6$ into the box at three different conditions (*top, middle, and bottom panels*). The plus mark and arrow indicate that the electric fans inside the box were switched on and the gas phase in the box was homogeneous, while the minus mark and arrow mean the fans were off. For details, see the text.

plus mark and arrow in the figure indicate that the fans were switched on and the gas in the box became homogeneous, while the minus mark and arrow mean that the fans were off.

There are three pairs of tracings in this figure for comparison. The top pair of curves, in which the tidal volume of 50 ml was smaller than the

58 ml of dead space volume, shows almost identical washins of helium and SF_6 into a well-stirred gas phase.

The bottom two curves show simultaneous washins of helium and SF_6 into a static gas phase, as indicated by the minus mark and arrow. In this case, the SF_6 washin seemed to be slower and of lesser magnitude than that of helium, but when the electric fans were switched on, as indicated by the plus mark and arrow, the concentration of SF_6 jumped up abruptly, while the helium concentration also increased, but to a lesser extent. Thus, SF_6 was transported into the box in an amount equal to or larger than the helium, but built up a larger concentration gradient, that is, exhibited stratification.

The middle two curves were obtained in a similar fashion, except that the ventilatory conditions were changed. The washin of indicator gases into the box depends on the speed of inspiration, tidal volume, and ventilatory frequency.

Stratification as a Function of Distance

In the next experiment, the indicator concentrations were measured as a function of distance, moving the tip of the sampling tube.

Figure 6.3 shows the results of simultaneous washin of 3% helium, argon, and SF_6, respectively, into a static gas phase. The abscissa indicates the distance between the distal end of the bronchial cast and the sampling site. Percent concentrations of the indicators relative to those in the inspirate are shown on the ordinate. Isotime lines at 20-sec intervals were drawn from a series of experiments. The result obtained (Fig. 6.3) shows that the heavier gas had a steeper decline of concentration toward the sampling site for a longer period of time. The concentrations of the indicator gases are in the order: SF_6, argon, and helium at the points close to the tip of the bronchial cast, but are reversed at the distance of 8 to 10 cm.

In analyzing the data, we have to take into consideration an interaction of convection and diffusion in the cnductive zone (Chang and Farhi 1973; Mazzone et al. 1976; Scherer et al. 1975; Wilson and LIn 1970), and uneven distribution of the inspired gas in the respective bronchi. Such an interaction, however, may favor a heavier gas over a lighter gas, and decrease the differences due to stratification.

In the actual lungs of the normal adult, a theoretical calculation on an anatomical basis predicts that the interface between 500 ml of inspired gas and the resident gas must lie approximately 2 mm from the alveolar wall in the absence of diffusive mixing. The situation is too far removed from the geometry of our model with a diffusing pathway 50 times longer to allow us to extrapolate from the present study to stratification in the human lungs under normal conditions. However, since binary diffusivity is

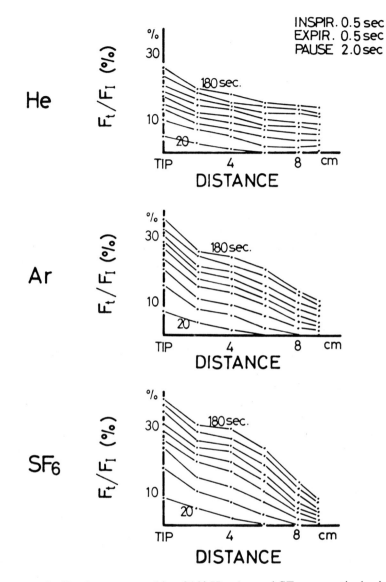

FIGURE 6.3. Simultaneous washin of 3% He, Ar, and SF$_6$, respectively, into the box, expressed as a function of the distance.

predicted to be inversely proportional to the ambient pressure (Paganelli and Kurata 1977), high pressure per se may be said to be equivalent, in a sense, to elongation of diffusion pathways in the lungs, and we presume that the present model may have something in common with some extreme situations leading to the stratification at depth.

Gas Exchange across the Alveolar-Capillary Membrane

As the second step, we attempted to quantitate the effects of stratification by measuring the time of appearance and the rate of transport of an inspired gas bolus into the arterial blood. Because of the complexity of blood oxygen and carbon dioxide transport and of the interaction between them, we resorted to using two inert gases of different diffusivities. Comparison of the behavior of two gases of essentially the same solubility obviates the necessity to know the exact distribution of gas flow, perfusion, and alveolar gas volume. Two such gases, used in a previous study by Adaro and Farhi (1971), are acetylene and monochlorodifluoromethane, which is also known as Freon-22. Molecular weights of these gases are 26 and 86.5, respectively.

The ratio of solubility of acetylene to that of Freon-22 in the blood of each animal was not identical, but was 1.06 on the average, which is sufficiently close for our purposes.

THE ANIMAL EXPERIMENTATION

Mongrel dogs were anesthetized, tracheostomized, and connected to a respirator through a three-way valve. A catheter for arterial blood sampling was introduced in the femoral artery and the free end of the catheter was connected to a manifold leading to several glass syringes, the dead space of which had been filled with a heparin solution.

At end-expiration, the animal's lungs were connected to an anesthesia bag in an airtight Lucite box through a valve. A preset volume of test gas of approximately 1.5% acetylene and Freon-22 in air was introduced into the lungs in 1 sec by applying pressure around the bag. Several arterial samples were obtained in rapid succession between 3 and 15 sec after inspiration of the test gas while the breath was held. Sampling always started with the most remote syringe, which was used for flushing the dead space and discarded. The single test breath volume was set at 300 ml in one series of runs and at 500 ml in another. Inspired gas and arterial blood samples were analyzed with a gas chromatograph (Carle Model 100 Micro-thermal Detector System) using a slight modification of the extraction technique described by Farhi et al. (1963).

Our working hypothesis was that the appearance of a more diffusible gas, acetylene, and of a less diffusible gas, Freon-22, into arterial blood during breath-holding would be similar to that shown schematically in Fig. 6.4. There should be relatively more acetylene in the early samples and relatively more Freon-22 in the later ones. Thus, the initial ratio of acetylene to Freon-22 in the early samples is expected to yield a value greater than 1.0 and decrease with time.

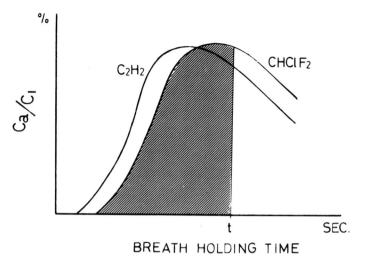

FIGURE 6.4. A working hypothesis on gas transport of two indicator gases across the alveolar-capillary membrane.

Fig. 6.5 shows a typical result of the time course of the two indicator gases. Back-extrapolation of the ascending limb to the abscissa yields an appearance time of approximately 3 sec, acetylene a bit earlier, consistent with our model experiment.

The concentration of acetylene was invariably higher than that of Freon-22 in the early samples, with a crossover point. The difference is small, but we must bear in mind that a considerable amount of smearing can take place, since it is likely that the inspired gas does not reach all areas of the lungs at the same time because of the asymmetric nature of the bronchial tree and because of a mixing in the blood as well.

Also, intra-acinar stratification of blood flow and gas concentration (Nixon and Pack 1980) may be a cause of smearing. All these factors would increase the differences.

In addition, axial penetration augmented by an interaction of convection and diffusion, if it occurs, will favor Freon-22 over acetylene and decrease the difference in appearance of the two gases in arterial blood. Consequently, it is probable that the acetylene-Freon relationships described here underestimate the actual differences between the two gases.

Fig. 6.6 indicates a typical result of the ratio of the two normalized gas fractions against time. The ratio is larger than 1.0 in the early samples and decreases with time, as mentioned before. The difference between the two test breath volumes of 300 ml and 500 ml is so small as to be of questionable significance, although the inspired gas volume has significant effect on the absolute values of the arterial contents of the tracers.

To allow a more meaningful comparison, the individual curves obtained

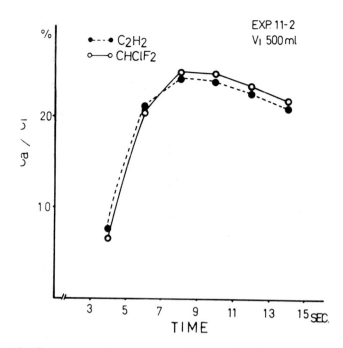

FIGURE 6.5. Time course of acetylene and Freon-22 in the arterial blood of a dog during breath-holding after inspiration of the test gas mixture. The test gas volume was 500 ml, and the breath was held at the end-inspiratory position.

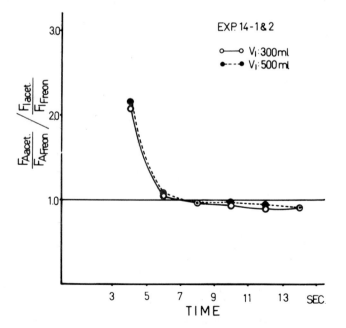

FIGURE 6.6. Time course of the ratio of acetylene in arterial blood to that of Freon-22 normalized by the fractions in the inspirate. The solid line indicates the experiment at the tidal test gas volume of 300 ml, while the dotted line means 500 ml.

in eight animals at the test gas volume of 500 ml were shifted horizontally, aligning each so as to get the crossover point, that is, the time at which alveolar fractions were equal, at time zero in Fig. 6.7.

Although there are some variations among the experiments, acetylene crosses the alveolar-capillary membrane faster than Freon-22, and the relative level of the two gases takes at least 4 sec to reach the equilibrium value in the lungs. Some of the variations in the equilibrium level can be ascribed to the slight differences in the ratios of the test gas volume to functional residual capacities.

DATA OBTAINED ON HUMANS

We extended the experiments to seven healthy human volunteers. Subjects inspired a test gas volume of 750 ml containing approximately 2% acetylene and Freon-22 in air, respectively, in 1 sec from FRC. The breath was held at either the end-inspiratory or end-expiratory position, and arterial samples were obtained.

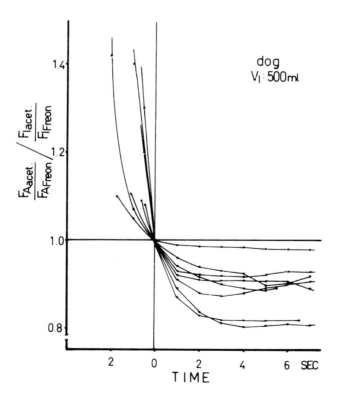

FIGURE 6.7. Time course of the ratio of acetylene in arterial blood to that of Freon-22. The individual curves obtained in eight dogs at the test gas volume of 500 ml were shifted horizontally, aligning each so as to get the crossover point at time zero.

The time course of the two indicators in arterial blood while the breath was held was virtually the same as that shown in Fig. 6.5 for the animal data. The difference between end-inspiratory and end-expiratory breath-holding was not significant, except for the maximum arterial contents of the tracers.

Fig. 6.8 shows results obtained from the experiments on the seven volunteers. As seen in the dog experiments (Fig. 6.7), the ratios of acetylene to Freon-22 were larger than 1.0 in the early samples and decreased with time. The equilibrium time in the lungs took at least 4 sec.

The exact mechanisms which separate the two tracers are not clear, but since the two gases vary in terms of molecular weight, and not solubility, the results obtained can be attributed to diffusion. Thus, the above-mentioned results strongly support the notion that diffusion within the gas phase in the lungs influences respiratory gas transport from the inspired gas to blood, especially at depths, while Van Liew et al. (1981) reported a minor effect of diffusion on gas exchange in hyperbaric environments.

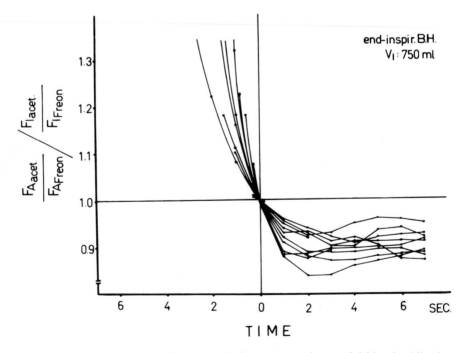

FIGURE 6.8. Time course of the two indicator gases in arterial blood while the breath was held at the end-inspiratory position. Ten records obtained in seven healthy volunteers were arranged in the same way as in Fig. 7.

Interpretation of the Chouteau Effect

We would now like to extend the analysis of the present studies one step further, and speculate on the effect reported by Chouteau and his associates (1969, 1971) in Marseilles, who described respiratory disturbances and hypoxemia in animals breathing normoxic or slightly hyperoxic gas mixtures at depths.

It is, of course, speculative to draw any definite conclusion from calculations heavily based on assumptions, but we feel it is legitimate to relate the experimental results, which we have reported, to the possible cause of the Chouteau effect.

The environmental pressures at which hypoxemic change was reported to appear in the simulated dives with four different gas mixtures were 91 ATA for He-O_2, 35 ATA for Ne-O_2, 21 ATA for N_2-O_2, and 16 ATA for Ar-O_2, respectively. It is clear that the animals breathing a lighter and therefore more diffusible gas mixture can endure a higher pressure in the environment without developing hypoxemia.

If we calculate binary diffusivities at the pressures at which the Chouteau effect was reported, using the Chapman-Enskog formula and assuming an inverse relationship between diffusivity and pressure (Paganelli and Kurata 1977), we find that hypoxemia occurred when the binary diffusivity was approximately 0.01 cm^2/sec, regardless of the gas species, as shown in Table 6.1. Table 6.2 indicates the relative time required for 90% diffusive gas mixing in the lungs as a whole, in reference to air at 1 atm. Numbers enclosed within squares indicate our calculated values for the relative mixing times required in the various environments at which the Chouteau effect was reported to appear. The values were quite similar to each other regardless of the gas species and the pressure.

In summary, although some of our conclusions are based on assumptions, our experimental work indicates that diffusion in the gas phase may be a limiting factor in gas transport at depths.

TABLE 6.1. Binary diffusion coefficients at depths calculated using the Chapman-Enskog formula.

	He-O_2	Ne-O_2	N_2-O_2	Ar-O_2
1 ATA	0.791	0.349	0.219	0.208
16 ATA	0.049	0.022	0.014	[0.013]
21 ATA	0.038	0.017	[0.010]	0.010
35 ATA	0.023	[0.010]	0.006	0.006
91 ATA	[0.009]	0.004	0.002	0.002

Numbers enclosed within squares indicate the binary diffusivities of the respective environmental gases at the pressures at which the Chouteau effect was reported to appear.

TABLE 6.2. Relative time required for 90% diffusive gas mixing in the lungs as a whole in reference to air at 1 atm.

	He-O_2	Ne-O_2	N_2-O_2	Ar-O_2
1 ATA	0.3	0.6	*1.0*	1.1
16 ATA	4	10	16	17
21 ATA	6	13	21	22
35 ATA	10	22	35	37
91 ATA	25	57	91	96

Numbers in squares indicate our calculated values for relative gas mixing times required in the various environments where the Chouteau effect was reported to appear.

REFERENCES

Adaro F, Farhi LE (1971). Effects of intralobular gas diffusion on alveolar gas exchange. *Fed Proc* 30: 437

Chang HK, Farhi LE (1973). On mathematical analysis of gas transport in the lung. *Respir Physiol* 18: 370–385

Chouteau J (1969). Saturation diving: The Conshelf experiments. In: Bennett PB, Elliott DH (eds) *The Physiology and Medicine of Diving*. Bailliere Tindall and Cassell, London, pp 491–504

Chouteau J, Guillerm R, Hee J, Pechon JCI (1971). Arterial hypoxia when breathing normoxic mixtures under hyperbaric conditions (abstract). XXXVth International Congress of Physiological Sciences Satellite Symposium; Recent Progress in Fundamental Physiology of Diving, Marseilles

Cumming G, Horsfield K, Jones JG, Muir DCF (1967). The influence of gaseous diffusion on the alveolar plateau at different lung volumes. *Respir Physiol* 2: 386–398

Farhi LE, Edwards AWT, Homma T (1963). Determination of dissolved N_2 in blood by gas chromatography and (a-A)N_2 difference. *J Appl Physiol* 18: 97–106

Georg J, Lassen NA, Mellemgaard K, Vinther A (1965). Diffusion in the gas phase of the lungs in normal and emphysematous subjects. *Clin Sci* 29: 525–532

Maio, DA, Farhi LE (1967). Effect of gas density on mechanics of breathing. *J Appl Physiol* 23: 687–693

Mazzone RW, Modell HI, Farhi LE (1976). Interaction of convection and diffusion in pulmonary gas transport. *Respir Physiol* 28: 217–225

Nixon W, Pack A (1980). Effect of altered gas diffusivity on alveolar gas exchange—A theoretical study. *J Appl Physiol* 48: 147–153

Okubo T, Piiper J (1974). Intrapulmonary gas mixing in excised dog lobes studied by simultaneous wash-out of two inert gases. *Respir Physiol* 21: 223–239

Paganelli CV, Kurata FK (1977). Diffusion of water vapor in binary and ternary gas mixtures at increased pressures. *Respir Physiol* 30: 15–26

Scherer PW, Shendalman LH, Greene NM, Bouhuys A (1975). Measurement of axial diffusivities in a model of the bronchial airways. *J Appl Physiol* 38: 719–723

Van Liew HD, Thalmann ED, Sponholtz DK (1981). Hindrance to diffusive gas mixing in the lungs in hyperbaric environments. *J Appl Physiol* 51: 243–247

Wilson TA, Lin KH (1970). Convection and diffusion in the airways and the design of the bronchial tree. In: Bouhuys A (ed) *Airway Dynamics*. Charles C Thomas, Springfield, IL pp 5–19

Worth H, Adaro F, Piiper J (1977). Penetration of inhaled He and SF_6 into alveolar space at low tidal volumes. *J Appl Physiol* 43: 403–408

7

Water Exchange in Hyperbaria

SUK KI HONG AND CHARLES V. PAGANELLI

Human divers are now able to engage in multiday dives to considerable depths using the mixed-gas saturation diving technique. In a dive recently carried out at Duke University, several divers reached a simulated depth of nearly 690 m (70 ATA) and then safely returned to sea level. Apparently, human divers seem able to cope with such high pressures and can perform normal activities albeit on a somewhat limited basis.

One of the major problems associated with hyperbaric exposure is the occurrence of high-pressure nervous syndrome (HPNS), which can severely limit the performance of divers. Fortunately, one can attenuate HPNS by slow compression or by use of a trimix (He-N_2-O_2) (Bennett and McLeod 1983). Another problem that has not been generally appreciated is the potential for body fluid disturbances at high pressure brought on by the development of a sustained diuresis.

A significant increase in urine flow was first observed by Hamilton et al. (1966) during a dry saturation dive conducted at a simulated depth of 650 ft (20.6 ATA). A similar hyperbaric diuresis was observed in many subsequent dives to various depths (Hong 1975). If the diuresis is sustained during a long dive, severe dehydration could develop, which will not only affect the performance of the divers but also endanger their safety.

The diuresis observed in many dives may be partly attributable to subtle cold stress. The thermoneutral temperature is considerably elevated in the hyperbaric environment (especially when He is used as a diluent gas) due to the high convective heat-transfer coefficient of dense gas mixtures (Webb 1970), and hence the environmental (hyperbaric chamber) temperature in many early saturation dives was below the thermoneutral level. However, in a limited number of dives, hyperbaric diuresis was still evident even in the complete absence of cold stress, indicating that hyperbaric exposure per se is responsible for the development of diuresis. The physiological characteristics of and the underlying mechanism(s) for this unique renal response to hyperbaric exposure will be presented in this review. Although there are many published papers

TABLE 7.1. Environmental parameters of three saturation dives conducted to study hyperbaric diuresis

Code (year)	Maximal pressure (ATA)	Bottom time at max. press. (days)	Temp. at press. (°C)	Rel. humidity at press. (%)	No. of subjects	Reference
Hana Kai II (1975)	18.6	17	30–31	70	5	Hong et al. (1977)
Seadragon IV (1979)	31	14	31.0±0.2	60±10	4	Nakayama et al. (1981)
Seadragon VI (1984)	31	7	31.5±0.5	60±10	4	Shiraki et al. (1987)

reporting the presence of hyperbaric diuresis, this review will be based on the data obtained from three dives (Table 7.1) in which comprehensive studies on body fluid balance were conducted.

General Characteristics of Hyperbaric Diuresis

Urine flow begins to increase during the compression phase, although it is not clearly known if there is a critical pressure at which the above response is elicited. Moreover, the possible relationship between the onset of diuresis and the compression rate is not defined. Urine flow continues to increase for several days after completion of compression, after which it declines somewhat but is still maintained at a higher-than-predive level throughout the rest of the hyperbaric period. During decompression, the above diuretic response is gradually reversed.

Fig. 7.1 shows that the increase in urine flow at high pressure is accompanied by significant reduction in urine osmolality but not by any change in the glomerular filtration rate (GFR) estimated by endogenous creatinine clearance (Ccr). The increase in daily urine flow at pressure is approximately 500 ml. Although urine osmolality always decreases at pressure, the osmolal clearance increased in all three dives listed in Table 7.1. This increase in osmolal clearance accounted for nearly 40–50% of the diuresis in Hana Kai II and Seadragon VI and for 100% in Seadragon IV. Thus negative free water clearance was reduced in the former two dives but remained unchanged in the latter. However, the ratio of negative free water clearance to osmolal clearance decreased 10–30% in all three dives, indicating that hyperbaric diuresis consists of both osmotic and free water components. These findings indicate that the primary mechanism for hyperbaric diuresis is inhibition of tubular reabsorption of osmotic particles as well as of free water, rather than increased GFR.

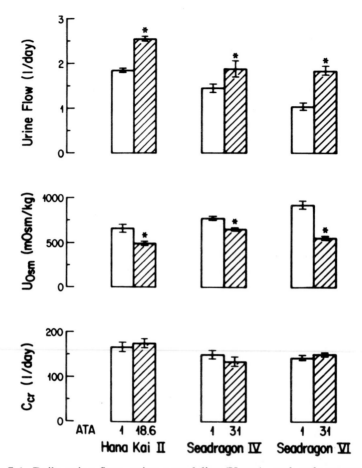

FIGURE 7.1. Daily urine flow, urine osmolality (Uosm), and endogenous creatinine clearance (Ccr) of human divers exposed to different pressures. Data for Hana Kai II, Seadragon IV, and Seadragon IV dives are obtained from Hong et al. (1977), Nakayama et al. (1981), and Shiraki et al. (1987), respectively. The symbol * indicates a value significantly different from the corresponding 1-ATA value (P < 0.05).

Despite the fact that diuresis was sustained throughout the period of hyperbaric exposure, daily total fluid intake (water drunk, water in food, and water of oxidation) decreased by approximately 10%, as shown in Fig. 7.2. This finding strongly suggests the development of significant dehydration. As stated above, the average increase in daily urine flow at pressure is ~500 ml/day. This means that, in the absence of any increase in daily water intake, net loss of body fluid during a 14-day hyperbaric exposure would amount to 7 kg, equivalent to 10% of body weight in a

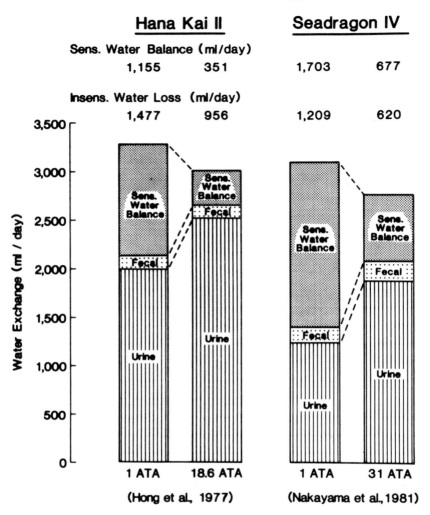

FIGURE 7.2. Daily water loss through various avenues under different pressures. The height of each bar represents the daily total fluid intake. [From Hong et al. (1983) with permission.]

70-kg diver. However, such a marked reduction in body weight was never observed in any of the three dives listed in Table 7.1. Body weight generally decreases slightly (0.5–1.0 kg) during the first 5 to 7 days of hyperbaric exposure, after which it gradually returns to the predive level (Hong et al. 1977; Matsuda et al. 1975; Nakayama et al. 1981). Moreover, both hematocrit and plasma protein concentration show significant increases only early in the hyperbaric period. Taken together, these findings indicate development of mild dehydration during the early phase of hyperbaric exposure, when the diuresis is greater than during the steady-state hyperbaric phase. Evidently, overall body fluid balance appears to be slightly disturbed only during the transition to hyperbaria for reasons that are entirely unknown at present. The notion that overall body fluid balance appears to be well maintained during the steady-state phase of hyperbaric exposure is supported by the fact that total body water volume, measured by the D_2O dilution technique at the midpoint of the Hana Kai II dive, was essentially the same as that determined before the dive (Hong et al. 1977).

Fig. 7.2 also shows the pattern of body water exchange through various avenues at 1-ATA air and in steady-state hyperbaric environments, assuming that overall body fluid balance is maintained even at pressure (see above). In the face of reduced daily water intake, it is clear that sensible water loss (urinary and fecal) is markedly increased at pressure in both dives. Consequently, sensible water balance (= total water input − sensible water loss) must be correspondingly decreased at pressure. In the two dives shown in Fig. 7.2, it decreased by 800–1000 ml/day at pressure. Sensible water balance represents the amount of body water lost from respiratory and skin surfaces via evaporative or insensible means. Since environmental temperature was maintained at around the thermoneutral range in these dives, insensible water loss should be the major component of the sensible water balance shown in Fig. 7.2. In fact, estimated daily insensible water loss, based on actual measurements during 3–5 hr before noon, decreased by 500–600 ml (Hong et al. 1977; Nakayama et al. 1981). The latter value is in general agreement with the average increase in daily urine flow at pressure. Thus, it appears that the primary mechanism for the free water component of hyperbaric diuresis is suppression of insensible water loss at pressure. Such a reduction in insensible water loss would result in a decrease in plasma osmolality and a subsequent inhibition of antidiuretic hormone (ADH) release, thereby inhibiting the tubular reabsorption of free water. In fact, plasma osmolality was found to decrease significantly at pressure in both Hana Kai II (Hong et al. 1977) and Seadragon VI (Shiraki et al. 1987). Moreover, urinary excretion of ADH has been shown to decrease significantly during the steady-state phase of hyperbaric exposure (Claybaugh et al. 1984; Hong et al. 1977; Claybaugh et al. 1987). More significantly, the plasma ADH level also showed a significant decrease at 31 ATA (Claybaugh et al. 1987). A theoretical basis for suppression of insensible water loss is presented below.

As stated above, hyperbaric diuresis also has an osmotic component. Increased urinary excretion of K, Pi (inorganic phosphate), urea, and Na contributes to the increase in osmolal clearance at pressure. Moreover, the osmotic component is more evident at night (see below). The mechanism for increase in excretion of these osmotic particles is not clearly understood.

Theoretical Basis for the Suppression of Skin Insensible Water Loss at Pressure

Viewed as a problem in mass transfer, the rate of water loss as vapor through the skin is directly related to the driving force on water vapor and inversely related to the total resistance through which the driving force acts. The driving force is the difference in concentration of water vapor between the tissues saturated at body temperature and the freely moving air beyond the adhering boundary layer (BL) of stagnant or slowly moving gas immediately adjacent to the skin surface. Water vapor diffusion occurs through the resistance offered by the skin and BL, and the water vapor is finally carried away by bulk gas flow. Thus, total resistance to water vapor flux is equal to the sum of skin resistance and BL resistance and can be described as follows (Davis et al. 1980):

$$E_c = \frac{\rho_{ts} - \rho_{ta}}{r_s + r_b} \tag{7.1}$$

where E_c = water loss per unit area through the skin ($g \cdot cm^{-2} \cdot sec^{-1}$), P_{ts} = saturation vapor density of water at body temperature ($g \cdot cm^{-3}$), P_{ta} = vapor density of water at ambient gas temperature ($g \cdot cm^{-3}$), r_s = skin resistance ($sec \cdot cm^{-1}$), and r_b = BL resistance ($sec \cdot cm^{-1}$) to water vapor flux. Rapp (1970) gives an analogous expression for evaporation from a wetted skin surface to ambient air, i.e., through the BL alone. In his treatment, BL resistance is shown to be directly related to the density of the gaseous medium (or to absolute pressure), as one would predict from the reciprocal relation between gas diffusivity and pressure given by the Chapman-Enskog equation (Reid et al. 1977).

Diffusion of water vapor through the BL will also depend on the composition of the gas phase in the boundary layer. In an atmosphere which is 98% He and 2% O_2, water vapor diffusivity is approximately three times greater than in air (Paganelli and Kurata 1977). In subjects exposed to such an environment at 18.6 ATA, one might expect skin insensible water loss to be reduced to 3/18.6, or about one-sixth of its value in air at 1 ATA. The data in Fig. 7.2 show a considerably smaller reduction, presumably because of unmeasured changes in other factors such as skin temperature, BL thickness (as a result of altered convective air flow), ambient water vapor density, and activity pattern of the subjects. In addition, systematic evaluation of skin resistance r_s to gas diffusion under hyperbaric conditions is lacking. There is evidence at 1 atm that r_s varies with environmental conditions such as relative humidity (Hale et al. 1958; Goodman and Wolf 1969; Oppermann and Heerd 1970). In the helium-rich environment usually found in hyperbaric chambers, one must take particular care that variables such as ambient humidity and skin temperature do not cause changes in r_s which can make interpretation of results uncertain at best.

Hyperbaric Nocturia

Urine flow displays a characteristic circadian rhythm in which peak flow is typically observed during the early afternoon and the lowest flow occurs at night. In order to assess the effect of hyperbaric exposure on such a circadian rhythm, the temporal pattern of urinary excretion of water and solutes was analyzed for the Seadragon IV and VI dives. Since the most interesting effect of high pressure was a nocturnal increase in urine flow, only diurnal and nocturnal data are shown in Fig. 7.3. At 18.6 ATA (Hana Kai II), urine flow increased similarly during day and night. However, at 31 ATA, urine flow showed a much greater increase during the night than during the day. In fact, urine flow did not even increase during the day in Seadragon IV. In both 31-ATA dives, the rate of endogenous creatinine excretion was the same during both day and night, indicating the absence of any circadian change in GFR. Hyperbaric nocturia appears to be a phenomenon that occurs at or above 25 ATA, at least for the Japanese divers involved in Seadragon IV and VI. Strangely enough, French divers exposed to 50 ATA did not show it (Rostain et al. 1975), while U.S. Navy divers exposed to 31 ATA showed a similar nocturia (personal communication from T. Doubt).

Regardless of possible ethnic differences in the development of hyperbaric nocturia, it is important to note that it is associated with a significant increase in osmolal clearance. As shown in Fig. 7.3, osmolal clearance at 31 ATA either decreased slightly or remained unchanged during the day but increased by 20–30% during the night. On the other hand, negative free water clearance at 31 ATA decreased by approximately 20% during the day but showed inconsistent changes at night. These results indicate that the hyperbaric diuresis observed at 31 ATA is largely the result of an increase in nocturnal urine flow resulting from an increase in osmolal clearance. On the other hand, a slight diurnal increase in urine flow at pressure appears to result from inhibition of free water reabsorption. This increase in free water excretion during the day at pressure is most likely the result of suppression of insensible water loss as described above. Naturally, free water excretion should also increase at night, but it is counteracted by increased osmolal clearance, which by itself increases negative free water clearance (Smith 1956). Indeed, urinary excretion of ADH at 31 ATA decreased in a similar manner during both day and night (Claybaugh et al. 1987).

Why does nocturnal excretion of osmotic particles increase at pressure, resulting in a hyperbaric nocturia? In order to answer this question, contributions of various solutes to the nocturnal increment in excretion of total osmotic substances at pressure were determined for the Seadragon VI dive (Shiraki et al. 1987). The urinary excretion of osmotic substances increased only during the night at 31 ATA by nearly 100 mosm (equivalent to ~10% of the daily excretion of total osmotic substances during the

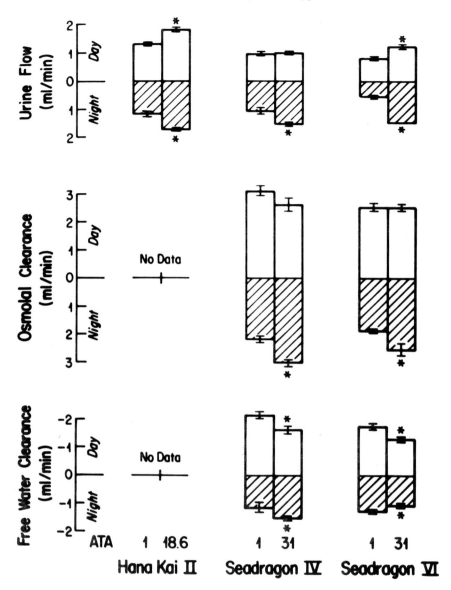

FIGURE 7.3. Diurnal (day) and nocturnal (night) urine flow, osmolal clearance, and free water clearance in the Hana Kai II, Seadragon IV, and Seadragon VI dives. See the legend for Figure 1 for references. The symbol * indicates a value significantly different from the corresponding 1-ATA value ($P < 0.05$).

predive period). The latter increment was contributed by urea (35%), NaCl (30%), K (15%), and others (20%). In this connection, it was interesting to note that urinary excretion of Na (and Cl) at 31 ATA actually decreased during the day but increased significantly at night.

Urinary excretion of urea and K at 31 ATA also showed a greater increase at night as compared with the daytime. It thus appears that there is nocturnal inhibition of tubular reabsorption of filtered Na at pressure. Such inhibition of Na reabsorption could secondarily increase excretion of urea, K, and water. Both plasma renin activity and aldosterone level increased significantly at 31 ATA. Urinary excretion of aldosterone also increased at pressure without showing any difference between day and night. In other words, the nocturnal increase in Na excretion was not accompanied by a greater inhibition of aldosterone level. Whatever the mechanism for the hyperbaric nocturnal natriuresis, it appears to be caused by a specific effect of high pressure on Na transport rather than a nonspecific effect on overall solute transport systems. This statement is supported by the fact that an increase in the excretion of Pi was observed only during the day and not at all at night.

Hormonal Changes Associated with Hyperbaric Diuresis

As discussed above, urinary excretion of ADH decreases significantly during the steady-state phase of hyperbaric exposure (Claybaugh et al. 1984; Claybaugh et al. 1987; Hong et al. 1977; Leach et al. 1978; Raymond et al. 1980). Moreover, plasma ADH level has also been shown to decrease significantly in Seadragon VI (Claybaugh et al. 1987). These results are consistent with the notion that the ADH system is suppressed during steady-state hyperbaric exposure, most likely in response to the suppression of insensible water loss (see above). The free water component of hyperbaric diuresis can certainly be explained on the basis of inhibition of the ADH system. However, a parallel relationship between changes in urine flow and ADH is not found during the early transitional period of hyperbaric exposure, when an increase in urine flow is accompanied by a significant increase in urinary excretion of ADH (Hong et al. 1977; Raymond et al. 1980). Moreover, this early hyperbaric diuresis is basically an osmotic diuresis leading to net dehydration, and hence it is not surprising that it develops through a mechanism not mediated by ADH. However, the latter mechanism is totally unknown at present. Another discrepancy between changes in urine flow and ADH has also been observed during the postdive 1-ATA control period. Typically, hyperbaric diuresis is totally reversed during the latter period, but urinary excretion of ADH either stays low (Claybaugh et al. 1984; Hong et al. 1977; Leach et al. 1978; Raymond et al. 1980) or shows partial recovery (Claybaugh et al. 1987). The plasma ADH level measured in Seadragon VI also remained lower during the postdive period than in the predive level (Claybaugh et al. 1987). The reason for this dissociation of the relationship between urine flow and ADH following long saturation dives is not obvious. It may suggest an increased sensitivity of the renal tubule to ADH. It is equally possible that heretofore unidentified factors may provide an antidiuretic force in the face of a continuous ADH suppression. Prostaglandin E_2 (PGE_2) is known to modulate tubular transport of water and Na. However, there were no significant changes in the urinary excretion of PGE_2 throughout the entire period (including the postdive phase) of Seadragon IV (Claybaugh et al. 1984). On the other hand, there was a twofold increase in the plasma level of prolactin during the postdive period as compared with the predive level (Claybaugh et al. 1984). Since prolactin is known to have an antidiuretic effect, the above finding may suggest a possible role of prolactin in salt and water balance following a prolonged dive.

Despite the increase in osmolal clearance at pressure, in part contributed by increased Na excretion (especially in Seadragon IV), the plasma renin activity, plasma aldosterone level, and urinary excretion of aldosterone showed consistent increases during and following hyyperbaric exposure (Claybaugh et al. 1984; Claybaugh et al. 1987; Hong et al. 1977). In other words, the increased excretion of Na at pressure is not due to an inhibition of the renin-aldosterone system. It appears that the increased renin-aldosterone level at pressure is a response to net dehydration induced by the early, transitional hyperbaric diuresis (see above). This means that there is another factor responsible for inhibition of Na reabsorption at pressure (see above). On the other hand, the increased excretion of K at pressure may be related to the increased renin-aldosterone level.

As discussed earlier, there were no obvious hormonal changes (ADH, renin-aldosterone) that could be correlated with the nocturnal increases in urine flow and excretion of certain osmotic particles (especially Na). It is not even known if hyperbaric nocturia is associated with the assumption of a supine (or prone) position or with the act of sleeping. Moreover, it is not known why phosphaturia is seen only during the day and not at night. The plasma level of parathyroid hormone (PTH) measured in early-morning samples showed no significant changes throughout the Seadragon VI dive (Claybaugh et al. 1987).

Transepithelial Na Transport under High Hydrostatic Pressure

As stated above, urinary excretion of Na (and the fractional excretion of filtered Na) increased, as did the renin-aldosterone level, during hyperbaric exposure (Claybaugh et al. 1984; Nakayama et al. 1981). This indicates the presence of a renin-aldosterone-independent mechanism (or mechanisms) which inhibits active tubular Na reabsorption under high pressure. One such factor may be high hydrostatic pressure itself. Indeed,

Goldinger et al. (1980) showed earlier that active Na efflux from human erythrocytes is inhibited reversibly by 30% at 30 ATA of hydrostatic pressure. More recent studies (Hong et al. 1984) also indicate that active Na transport across toad skin is inhibited by high pressure. The toad skin preparation is widely used for studies on Na transport since it is functionally analogous to the distal tubule of the mammalian kidney. Short-circuit current (Isc) which can be relatively easily measured under high pressure was first validated as a measure of active (net) Na transport at high pressure by Goldinger et al. (1986). Upon compression, baseline Isc increased during the first 10 min (by 20%) and then decreased continuously, leveling off after 30–40 min at pressure. This steady-state effect of pressure was related to the level of pressure in a curvilinear manner (~20% at 50 ATA), leveling off at 200 ATA, and the overall pressure-response relationship was not much different from that for active Na efflux from human erythrocytes. These results strongly support the possibility that active transtubular Na reabsorption may also be subject to inhibition under high pressure and thus lead to hyperbaric natriuresis. According to the current cellular model for active Na transport across toad skin, entry of Na into the epithelial cell across the outer (or apical) membrane is passive, while extrusion of Na from the cell across the inner (or basolateral) membrane is active and is mediated by the Na-K-ATPase (Macknight et al. 1980). Therefore, the inhibition of epithelial Na transport observed under high pressure could be explained by reduction of apical membrane Na permeability and/or inhibition of the Na-K exchange pump in the basolateral membrane.

Studies on toad skin have also provided some indirect evidence to indicate that high hydrostatic pressure (200–300 ATA) interferes with one of the steps involved in the formation of cAMP (the intracellular mediator of ADH) (Hong et al. 1984). It will be interesting to see if such an interference with ADH-induced cAMP production could also take place at high pressure in the mammalian kidney. This information would shed light on the mechanism of the possible alteration of the ADH action at pressure.

Summary

Urine flow increases during prolonged exposure of human divers to a hyperbaric environment. Typically, this hyperbaric diuresis develops in the absence of any increase in either GFR, daily fluid intake or cold stress. Comprehensive fluid balance studies show that a significant reduction of insensible water loss in hyperbaria is primarily responsible for the diuresis. This view is also supported by the fact that hyperbaric diuresis is accompanied by an increase in free water excretion and significant decreases in plasma ADH level and urinary excretion of ADH.

In vitro experiments also demonstrate an inverse relationship between diffusivity of water vapor and pressure (or gas density). When environmental pressure exceeds 25 ATA, however, the diuresis is characterized by an increase in osmolal clearance, especially during the night. Such a nocturnal diuresis is not accompanied by any change in the pattern of urinary excretion of ADH or aldosterone. In vitro studies indicate that active Na transport across an epithelium (toad skin) is indeed inhibited by high hydrostatic pressure, an observation which may at least in part account for the osmotic component of the diuresis observed in hyperbaria.

REFERENCES

Bennett PB, McLeod M (1983). Comparative effect of compression rate and trimix (He/N_2/O_2) on performance at depth to 686 m. In: Shiraki K, Matsuoka S (eds) *Hyperbaric Medicine and Underwater Physiology.* Program Committee of III UOEH Symposium, Kitakyushu, Japan, pp 179–188

Claybaugh JR, Hong SK, Matsui N, Nakayama H, Park YS, Matsuda M (1984). Responses to salt and water regulating hormones during a saturation dive to 31 ATA (Seadragon IV). *Undersea Biomed Res* 11: 65–80

Claybaugh JR, Matsui N, Hong SK, Park YS, Nakayama H, Shiraki K, Seadragon VI (1987). A 7-day dry saturation dive at 31 ATA. III. Alterations in basal and circadian endocrinology. *Undersea Biomed Res* 14: 401–412

Davis JE, Spotila JR, Schefler WC (1980). Evaporative water loss from the American alligator, *Alligator Mississippiensis:* The relative importance of respiratory and cutaneous components of the regulatory role of the skin. *Comp Biochem Physiol* 67A: 439–466

Goldinger JM, Duffey ME, Morin RA, Hong SK (1986). The ionic basis of short-circuit current in toad skin at elevated hydrostatic pressure. *Undersea Biomed Res* 13: 361–367

Goldinger JM, Kang BS, Chou YE, Paganelli CV, Hong SK (1980). Effect of hydrostatic pressure on ion transport and metabolism in human erythrocytes. *J Appl Physiol* 49: 224–231

Goodman AB, Wolf AV (1969). Insensible water loss from human skin as a function of ambient vapor concentration. *J Appl Physiol* 26: 203–207

Hale FC, Westland RA, Taylor CL (1958). Barometric and vapor pressure influences on insensible loss. *J Appl Physiol* 12: 20–28

Hamilton RW, MacInnis JB, Noble AD, Schreiner HR (1966). Saturation diving to 650 feet. Tech Memorandum B-411, Ocean Systems, Inc., Tonawanda, NY

Hong SK (1975). Body fluid balance during saturation diving. In: Hong SK (ed) *International Symposium on Man in the Sea.* Undersea Medical Society, Bethesda, pp 127–140

Hong SK, Claybaugh JR, Frattali V, Johnson R, Kurata F, Matsuda M, McDonough AA, Paganelli CV, Smith RM, Webb P (1977). Hana Kai II: A 17-day dry saturation dive at 18.6 ATA. IV. Body fluid balance. *Undersea Biomed Res* 4: 247–265

Hong SK, Claybaugh JR, Shiraki K (1983). Body fluid balance in the high pressure environment. In: Shiraki K, Matsuoka S (eds) *Hyperbaric Medicine and Underwater Physiology*. Program Committee of III UOEH Symposium, Kitakyushu, Japan, pp 223–234

Hong SK, Duffey ME, Goldinger JM (1984). Effect of high hydrostatic pressure on sodium transport across the toad skin. *Undersea Biomed Res* 11: 37–47

Leach CS, Cowley JRM, Troell MT, Clark JM, Lambertsen CJ (1978). Biochemical, endocrinological and hematological studies. In: Lambertsen CJ, Gelfand R, Clark JM (eds) *Predictive Studies IV, Work Capacity and Physiological Effects in He-O₂ Excursions to Pressures of 400–800–1200 and 1600 Feet of Sea Water*. University of Pennsylvania, Philadelphia, pp E17.1–59

Macknight ADC, DiBona DR, Leaf A (1980). Sodium transport across toad urinary bladder: A model "tight" epithelium. *Physiol Rev* 60: 615–715

Matsuda M, Nakayama H, Kurata FK, Claybaugh JR, Hong SK (1975). Physiology of man during a 10-day dry heliox saturation dive to 7 ATA. II. Urinary water, electrolytes, ADH and aldosterone. *Undersea Biomed Res* 2: 119–131

Nakayama H, Hong SK, Claybaugh J, Matsui N, Park YS, Ohta Y, Shiraki K, Matsuda M (1981). Energy and body fluid balance during a 14-day dry saturation dive at 31 ATA (Seadragon IV). In: Bachrach JA, Matzen MM (eds) *Underwater Physiology VII. Proceedings of the 7th Symposium on Underwater Physiology*. Undersea Medical Society, Bethesda, MD, pp 541–554

Oppermann C, Heerd E (1970). Die Wasserdampfdiffusion durch die menschliche Epidermis *in vivo* unter der Einwirkung unterschiedlicher Umweltfeuchte und mach Wasserim bibition. *Pflügers Arch* 318: 51–62

Paganelli CV, Kurata FK (1977). Diffusion of water vapor in binary and ternary gas mixtures at increased pressures. *Respir Physiol* 30: 15–26

Rapp GM (1970). Convective mass transfer and the coefficient of evaporative heat loss from human skin. In: Hardy JD, Gagge AP, Stolwijk JAJ (eds) *Physiological and Behavioral Temperature Regulation*. Charles C Thomas, Springfield, IL, pp 55–80

Raymond LW, Raymond NS, Frattali VP, Sode J, Leach CS, Spaur WH (1980). Is the weight loss of hyperbaric habituation a disorder of osmoregulation? *Aviat Space Environ Med* 51: 397–401

Reid RC, Prausnitz JM, Sherwood TK (1977). *Properties of Gases and Liquids*, 3rd ed. McGraw-Hill, New York

Rostain JC, Naquet R, Reinberg A (1975). Effects of hyperbaric saturation (500 meter depth, heliox atmosphere) on circadian rhythms of two healthy young men. *Internat J Chronobiol* 3: 127–139

Shiraki K, Hong SK, Park YS, Sagawa S, Konda N, Claybaugh JR, Takeuchi H, Matsui N, Nakayama H (1987). Seadragon VI: A 7-day dry saturation dive at 31 ATA. II. Characteristics of diuresis and nocturia. *Undersea Biomed Res* 14: 387–400

Smith W (1956). *Principles of Renal Physiology*. Oxford University Press, New York, pp 120–124

Webb P (1970). Body heat loss in undersea gaseous environments. *Aerosp Med* 41: 1282–1288

8

On the Use of a Bubble Formation Model to Calculate Nitrogen and Helium Diving Tables

D.E. YOUNT AND D.C. HOFFMAN

Introduction

Decompression sickness is caused by a reduction in ambient pressure which results in supersaturation and the formation of gas bubbles in blood or tissue. This well-known disease syndrome, often called "the bends," is associated with such modern-day activities as deep-sea diving, working in pressurized tunnels and caissons, flying at high altitudes in unpressurized aircraft, and flying EVA excursions from spacecraft. A striking feature is that almost any body part, organ, or fluid can be affected, including skin, muscle, brain and nervous tissue, the vitreous humor of the eye, tendon sheath, and bone. Medical signs and symptoms range from itching and mild tingling sensations to crippling bone necrosis, permanent paralysis, and death.

The generality of the symptoms of decompression sickness and the fact that humans consist mainly of water suggest that the problem of bubble formation in the body may have a simple physical solution. Furthermore, since bubble formation occurs in almost any aqueous medium, the phenomenon can be studied in whatever substance is most convenient. A frequent choice in the series of experiments carried out at the University of Hawaii has been unflavored Knox gelatin, which is transparent and holds bubbles in place so that they can be counted and measured (Strauss 1974; Strauss and Kunkle 1974; Yount and Strauss 1976; Yount et al. 1979; Yount and Yeung 1981). Distilled water, seawater, agarose gelatin (D'Arrigo 1978), and infertile hen's eggs (Paganelli et al. 1977) have also been used.

The main outcome of this line of investigation has been the development of the varying-permeability model (VPM), in which cavitation nuclei consist of spherical gas phases that are small enough to remain in solution and strong enough to resist collapse, their stability being provided by elastic skins or membranes consisting of surface-active molecules (Yount 1979b). Ordinarily VPM skins are permeable to gas, but they can become effectively impermeable when subjected to large compressions, typically exceeding 8 atm.

By tracking the changes in the nuclear radius that are caused by increases or decreases in ambient pressure, the varying-permeability model has provided precise quantitative descriptions of several of the bubble-counting experiments carried out in supersaturated gelatin (Yount and Strauss 1976; Yount et al. 1979; Yount and Yeung 1981). The model has also been used to trace levels of incidence for decompression sickness in a variety of animal species, including salmon, rats, and humans (Yount 1979a, 1981). Finally, microscopic evidence has recently been obtained (Yount et al. 1984) which indicates that spherical gas nuclei—the persistent microbubbles hypothesized by the varying-permeability model— actually do exist and have physical properties consistent with those previously assigned to them (Yount 1979b; Yount and Strauss 1976; Yount and Yeung 1981; Yount et al. 1979). Nuclear radii, for example, are on the order of 1 μm or less, and the number density decreases exponentially with increasing radius (Yount 1979b; Yount and Strauss 1976; Young and Yeung 1981; Yount et al. 1979). The exponential radial distribution is believed to be characteristic of a system of VPM nuclei in thermodynamic equilibrium, and it can be derived from statistical-mechanical considerations (Yount 1982).

The most recent step in applying the varying-permeability model to decompression sickness has been to calculate a comprehensive set of nitrogen and helium diving tables and compare them with other tables now in use. The work on nitrogen has been reported already at scientific meetings (Yount and Hoffman 1983, 1984) and in a journal article (Yount and Hoffman 1986); the results for helium are being given for the first time here.

The Decompression Criterion

Early applications of the varying-permeability model to decompression sickness (Yount 1979a, 1981) involved rudimentary pressure schedules in which the subjects were first saturated with gas at some elevated pressure P_1 and then supersaturated by reducing the pressure from P_1 to the final setting P_2. The data in such experiments are most easily presented by plotting the combinations of supersaturation vs. exposure pressure ($P_{ss} \simeq P_1 - P_2$ vs. P_1), which yield a given morbidity, for example, a 50% probability of contracting decompression sickness. In order to describe these data, it was assumed that lines of constant morbidity were also lines of constant bubble number N (Yount 1979a, 1981). The bubble number, in turn, was assumed to be equal to the number of spherical gas nuclei with initial radii r_0 larger than some minimum radius r_0^{min} (Yount 1979b). This approach was remarkably successful, partly because the schedules involved were so simple—representing, as it were, a type of controlled experiment in which most of the variables in the problem were fixed.

The naive assumption of constant nucleation or constant bubble number does not encompass the full range of conditions covered by modern diving tables. That is, it yields a set of tables which, though they may be very safe, do not track conventional tables in their global behavior and often require total ascent times that would generally be considered excessive by the commercial diving industry. Given these circumstances, a decision was made to treat the conventional tables as valid experimental data and reformulate the decompression criterion accordingly.

The first step was to replace "constant bubble number" with a "critical-volume hypothesis," thereby assuming that signs or symptoms will appear whenever the total volume V accumulated in the gas phase exceeds some designated critical value V_{crit}. Although V_{crit} itself is fixed for all of the diving tables, gas is continuously entering and leaving the gas phase. In this sense, the new decompression criterion is dynamic, rather than static as in other applications of the critical-volume point of view (Hennessy and Hempleman 1977).

The idea that gas is continuously leaving the gas phase is suggested by previous applications (Yount 1979a, 1981), which seem to imply that there is a bubble number N_{safe} which can be tolerated indefinitely, regardless of the degree of supersaturation P_{ss}. From this, it was deduced that the body must be able to dissipate free gas at a useful rate that is proportional both to N_{safe} and to P_{ss} (Yount and Hoffman 1986). A possible rationale is provided by physiological studies which demonstrate that, so long as its capacity is not exceeded, the lung is able to continue functioning as a trap for venous bubbles (Butler and Hills 1979).

Another implication of the present investigation is that in practical diving tables (and especially in surface-decompression procedures), the actual number of supercritical nuclei N_{actual} is allowed temporarily to exceed the number which can be tolerated indefinitely N_{safe}. This permits the volume of the gas phase to inflate at a rate that is proportional to P_{ss} ($N_{actual} - N_{safe}$). In the present formulation, the increase in gas-phase volume continues until P_{ss} is zero. At this point, usually long after the dive has ended, the net volume of released gas has reached its maximum value V_{max}, which must be less than V_{crit} if signs and symptoms of decompression sickness are to be avoided.

Computation of a Diving Table

Computation of a diving table (Yount and Hoffman 1986) begins with the specification of six nucleation parameters. These are the pressure P* at which the nuclear skins become impermeable to gas, the surface tension γ, the nuclear skin compression γ_C, the time constant τ_R for the regeneration of nuclei crushed in the initial compression, and a composite parameter λ which is related to V_{crit} and determines, in effect, the amount by which the actual bubble number N_{actual} can exceed the safe bubble number N_{safe}. N_{actual} is much larger than N_{safe} for short dives, but the two are nearly equal for dives of long duration.

From the given set of parameter values, the program calculates a preliminary estimate of P_{ss} that is just sufficient to probe the minimum initial radius r_0^{min} and hence to produce a number of bubbles equal to N_{safe}. In the permeable region of the model, which includes the great majority of air diving tables, the nuclear radius r_1^{min} following an increase in pressure from P_0 to P_1 can be obtained from the equation (Yount 1979b)

$$(1/r_1^{min}) = (1/r_0^{min})\,(P_1 - P_0)/2(\gamma_C - \gamma) \tag{8-1}$$

Regeneration of the nuclear radius is allowed to take place throughout the time t_R during which the pressure is held at P_1. Regeneration may occur through a complex statistical-mechanical process (Yount 1982), which is approximated here via an exponential decay with regeneration time constant τ_R:

$$r(t_R) = r_1^{min} + (r_0^{min} - r_1^{min})\,[1 - \exp(-t_R/\tau_R)] \tag{8-2}$$

The supersaturation P_{ss}^{min} that is just sufficient to probe r_0^{min} is then found from (Yount 1979b)

$$P_{ss}^{min} = 2(\gamma/\gamma_C)\,(\gamma_C - \gamma)/r(t_R) \tag{8-3}$$

Holding P_{ss}^{min} fixed, the program next calculates a decompression profile and the total decompression time t_D. From t_D a new value of P_{ss}^{min} is obtained which will probe a new initial radius r_0^{new} that is smaller than r_0^{min} and hence will result in a number of bubbles that is larger than N_{safe}. The relevant equation is (Yount and Hoffman 1986)

$$P_{ss}^{new} = [b + (b^2 - 4c)^{1/2}]/2 \tag{8-4a}$$

where

$$b = P_{ss}^{min} + \lambda\gamma[\gamma_C(t_D + H/0.693)] \tag{8-4b}$$

and

$$c = (\gamma/\gamma_C)^2\lambda(P_1 - P_0)/(t_D + H/0.693) \tag{8-4c}$$

Using the respective values of P_{ss}^{new} for each tissue half-time H, the program determines a more severe decompression profile, which will yield updated values of t_D and P_{ss}^{new}. After several iterations, t_D and P_{ss}^{new} converge, implying that V_{max} now differs from V_{crit} by an acceptably small amount.

The uptake and elimination of inert gas by the body are assumed to be exponential, as in conventional tables. Water vapor pressure and the dissolved partial pressures of oxygen and carbon dioxide are calculated in the manner described by Yount and Lally (1980). The net contribution of these "active" gases is nearly constant at 102 mm Hg for inspired oxygen pressures up to about 2 atm abs (Yount and Lally 1980). This limit is not reached for air decompression tables at ambient pressures below about 10 atm abs. In the case of air or other nitrogen-oxygen mixtures, the half-times H for the various "tissue compartments" are nominally 1, 2, 5, 10, 20, 40, 80, 120, 160, 240, 320, 400, 480, 560, and 720 min, but the 1-min tissue is never used. The half-times for helium were obtained by dividing each of the nitrogen half-times by 3 (Lambertsen and Bardin 1973).

Setting the Nucleation Parameters

For the original air diving tables (Yount and Hoffman 1986), the imperme-ability parameter P* was assigned a value of 9.2 atm abs based on an analysis (Yount 1979b) of experiments carried out in Knox gelatin (Yount and Strauss 1976). This is equivalent to about 270 fsw (33 fsw=10 msw=1 atm=2 atm abs, etc.), which is near the 300-fsw upper limit of those original calculations and well above the level at which nitrogen narcosis and oxygen toxicity begin to be concerns. With this choice of P*, nearly all of the original tables were in the "permeable" or "linear" region described by Eqs. (1) and (3), and it was suggested that P* and the associated impermeable equations could be deleted altogether (Yount and Hoffman 1986).

The initial attempts to extend the VP model to helium were made with P* still equal to 9.2 atm abs. For excursions beyond 200 fsw, it appeared that the resulting decompression times might be too short. The air diving tables can accommodate a lower P*; hence the final setting for both nitrogen and helium is 5.0 atm abs. The lower value for P* implies that decompression times are now somewhat longer and more conservative for all of the deeper dives.

Since the model predictions depend only upon the ratios γ/γ_C and $2\gamma/r_0^{min}$, the setting of γ is essentially arbitrary (Yount 1979a, 1981). To be definite, however, a value of $\gamma = 17.9$ dyn/cm was selected (Davson 1964). With this choice, the values of the remaining four parameters for nitrogen are $\gamma_C = 257$ dyn/cm, $r_0^{min} = 0.8$ μm, $\tau_R = 20,160$ min, and $\lambda = 7500$ fsw-min. These were found by requiring that the total decompression times of the new tables resemble those of the Tektite saturation dive (Beckman and Smith 1972) and of the U.S. Navy (1970) and Royal Naval Physiological Laboratory (1968) manuals.

In the extrapolation from nitrogen to helium, most of the nucleation parameters should remain the same. It has already been noted that P* is now equal to 5.0 atm abs for both gases. Similarly, the surface tension γ, nuclear skin compression γ_C, regeneration time constant τ_R, and critical volume of released gas V_{crit} should not be altered. If V_{crit} is fixed, then the value of the composite parameter λ is also constant. The reason there is no change in τ_R is that regeneration of VP nuclei is a stochastic process which depends upon the ambient temperature and certain properties of the skin and is independent of the choice of inert gas (Yount 1982). The only nucleation parameter left is r_0^{min}.

It has been observed both in Knox gelatin (Yount 1978) and in agarose gelatin (D'Arrigo 1978) that the number of visible bubbles is similar for various gases subjected to a given pressure schedule. Bubble size, however, depends upon the solubility. In effect, each supercritical nucleus collects the excess dissolved gas in its immediate vicinity, competing with neighboring nuclei until equilibrium is reached and the dissolved gas tension matches the ambient pressure.

To obtain a predetermined volume V_{crit} with the smaller helium bubbles, the bubble number must be higher. To induce a larger bubble number, the supersaturation pressures P_{ss}^{min} and P_{ss}^{new} must be increased, thereby reducing the value of the critical radius r_0^{min}. The empirical result for helium is 0.7 μm, only slightly lower than the 0.8 μm found for nitrogen.

One other factor must be taken into account in going from nitrogen to helium, and this is the counterdiffusion phenomenon (Lambertsen and Idicula 1975). Whereas divers subjected to elevated pressures of air or nitrogen-oxygen mixtures are breathing the same inert gas prior to exposure, those subjected to elevated pressures of helium-oxygen ordinarily are not. Helium divers must therefore endure not only a pressure excursion, but also a switch from one inert gas to another. While the helium is diffusing into the various tissue compartments at a rate that is relatively fast, residual nitrogen is diffusing out at a rate that is relatively slow. This is reflected in the tissue half-times H, which were assumed to be in the ratio of 1 to 3 for the two gases, as already noted. Counterdiffusion can result in supersaturation even if there is no change in ambient pressure (Lambertsen and Idicula 1975), and it can enhance the degree of supersaturation during pressure excursions.

To summarize this section, all of the results reported in this paper were obtained by optimizing the values of five adjustable nucleation parameters, P^*, γ_C, r_0^{min}, τ_R, and λ, which replace the U.S. Navy's matrix of M-values. The impermeability threshold P^* comes into play only for the deeper dives, and the regeneration time constant τ_R has a significant effect only for saturation dives. For the great majority of depths and durations that divers actually experience, three nucleation parameters would suffice. The only parameter which is different for helium and nitrogen is the critical radius r_0^{min}, which is 0.7 μm in the former case and 0.8 μm in the latter.

Depths and pressures are given in feet of seawater for convenience in making comparisons with the Tektite, U.S. Navy (USN), and Royal Naval Physiological Laboratory (RNPL) reference schedules. For similar total decompression times, the set of tables generated in this study is expected to yield smaller total bubble volumes and therefore to be safer. However, none of the tables has yet been tested on either animal or human subjects.

Results for Nitrogen

In this section, the salient features of a number of diving tables using oxygen-nitrogen as the breathing gas mixture are reviewed. Further details can be found in the paper by Yount and Hoffman (1986). The VPM, USN, and RNPL profiles for an "exceptional exposure" involving greater-than-normal risk are given in Table 8.1. The descent and ascent rates are 60 fsw/min, and the 3.33 min required to reach 200 fsw is counted as part of the 60-min "bottom time." The total decompression times are similar for the VPM and USN, the important difference being the deeper "first stop" of the VPM schedule, 130 fsw vs. 60 fsw for the USN. The first stop for the RNPL is at 100 fsw, somewhat deeper than the USN, but still well below the VPM. Deeper first stops are a persistent feature of the literally hundreds of VPM schedules that have been compared with conventional tables now in use. The calculations indicate that the longer "first pull" of these conventional tables results in a larger supersaturation P_{ss}, in a larger bubble number N, and ultimately in a larger maximum volume of released gas V_{max}.

VPM, USN, and RNPL "no-stop" decompressions are compared in Table 8.2. As in Table 8.1, the breathing gas is air. The absence of prolonged decompression stages makes these types of "data" nearly independent of the overall surfacing strategy. The allowed times for the VPM are everywhere very similar to those for the RNPL, and the VPM and USN agree exactly at 5 min, 180 fsw, and again at 5 min, 190 fsw.

TABLE 8.1. Comparing the VPM, USN, and RNPL, air decompression schedules for a 60-min, 200-fsw excursion.

Stop depth (fsw)	VPM wait (min)	USN wait (min)	RNPL wait (min)
130	1		
120	3		
110	3		
100	4		5
90	7		0
80	6		0
70	7		5
60	11	2	5
50	16	13	10
40	17	17	35
30	22	24	45
20	39	51	180
10	52	89	2*
Total ascent time	191:20	199:20	289:00

* No wait at 10 fsw; rate of ascent from 20 fsw is 10 fsw min.

TABLE 8.2. Comparing the no-decompression limits of the VPM, RNPL, and USN air tables.

Allowed depth (fsw)	Source table type	Allowed time (min)	Allowed depth (fsw)	Source table type	Allowed time (min)
30	VPM	323	140	VPM	8
	RNPL	Unlimited		RNPL	7.5
	USN	Unlimited		USN	10
40	VPM	108	150	VPM	7
	RNPL	102		RNPL	Not listed
	USN	200		USN	5
50	VPM	63	160	VPM	7
	RNPL	60		RNPL	5.5
	USN	100		USN	5
60	VPM	39	170	VPM	6
	RNPL	41		RNPL	Not listed
	USN	60		USN	5
70	VPM	30	180	VPM	5
	RNPL	30		RNPL	4.5
	USN	50		USN	5
80	VPM	23	190	VPM	5
	RNPL	23		RNPL	Not listed
	USN	40		USN	5
90	VPM	18	200	VPM	4
	RNPL	18		RNPL	3.5
	USN	30		USN	Not allowed
100	VPM	15	210	VPM	4
	RNPL	15		RNPL	Not listed
	USN	25		USN	Not allowed
110	VPM	12	220	VPM	4
	RNPL	Not listed		RNPL	3
	USN	20		USN	Not allowed
120	VPM	11	230	VPM	4
	RNPL	10		RNPL	Not listed
	USN	15		USN	Not allowed
130	VPM	10	240	VPM	Not allowed
	RNPL	Not listed		RNPL	Not allowed
	USN	10		USN	Not allowed

Over the entire range, the VPM appears to provide a safe, tight, and therefore useful lower bound. The fact that the VPM is systematically lower than the USN at the shallower depths reflects the general conservatism of the tables as a whole. A bolder, more aggressive set of tables could, of course, be computed by simply adjusting the values of the nucleation parameters.

One very practical reason for attempting to optimize decompression procedures from first principles is the hope that if a correct global theory can someday be formulated, it will be possible to relate and describe the whole range of decompression experience with a small number of equations and parameter values. Instead of "titrating" a handful of "volunteers" to develop a new table or determine a new M-value, a method which necessarily has limited statistical accuracy, one will be able to use an already-calibrated theory to interpolate or extrapolate, thereby bringing to bear the full statistical weight of a much larger data base.

The global potential of the nucleation approach can be illustrated by comparing an appropriate VPM schedule with the profile actually used by humans without incident on the 14-day, 100-fsw Tektite saturation dive (Beckman and Smith 1972). Because the actual breathing gas was a normoxic oxygen-nitrogen mixture rather than air, the VPM schedule was calculated for a 14-day exposure to the 126-fsw equivalent air depth (Yount and Lally 1980) of the Tektite dive. The actual Tektite decompression required 2960 min of breathing the normoxic mixture, plus an additional 170 min of breathing pure oxygen. Because oxygen facilitates nitrogen washout, the 170 min during which pure oxygen was used was equivalent to approximately 340 min of breathing the normoxic mixture. Had pure oxygen not been used, therefore, the actual decompression time would have been around 3300 min. The result for the VPM is similar at 3530 min.

Results for Helium

The VPM no-stop decompressions for helium are compared with those of the U.S. Navy (Flynn et al. 1981; Schilling et al. 1976) in Fig. 8.1. "VPM Helium" refers to the unusual case in which the diver has equilibrated on helium before the exposure to elevated pressure, and "VPM Residual Nitrogen" refers to the more common situation in which the diver is initially breathing air and is then subjected both to elevated pressure and to a switch from nitrogen to helium.

As discussed previously, counterdiffusion of the two inert gases when breathed in the order nitrogen-to-helium can enhance the degree of supersaturation produced during a pressure excursion (Lambertsen and Idicula 1975). The result is that the curve for VPM residual nitrogen is lower and more conservative than that for VPM helium. Both VPM curves lie below the USN curve at the shallower depths, but the results are similar near 200 fsw. This parallels the comparison made earlier in Table 8.2, where the VPM agreed well with the RNPL over the entire range but was much more conservative than the USN except at the greatest depths.

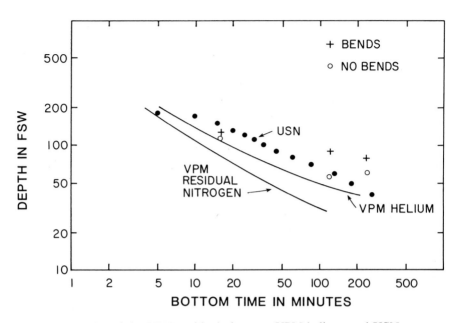

FIGURE 8.1. Plot of the VPM residual nitrogen, VPM helium, and USN no-stop decompression limits. Several data points are included for reference.

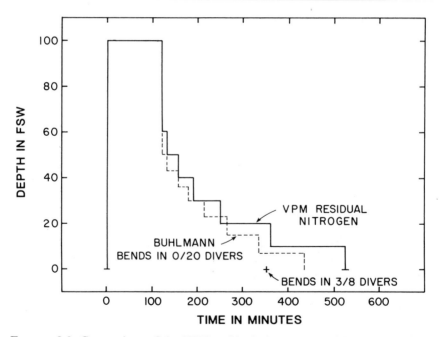

FIGURE 8.2. Comparison of the VPM residual nitrogen and Bühlmann schedules for a 120-min, 100-fsw excursion. The profiles are quite similar, but the VPM total time is farther from the bends threshold.

The VPM residual nitrogen schedule plotted in Fig. 8.2 is for a depth of 100 fsw and a bottom time of 120 min breathing 79% helium and 21% oxygen. Also shown is an experimental dive developed by Bühlmann (1984). Bühlmann's "titration" method involved shortening the last stop of a comparatively safe decompression until bends symptoms were observed. The two curves are remarkably similar except for this last stop. Bühlmann found that a total decompression time of 316 min (total exposure of 436 min) resulted in no bends in 20 subjects, but 232 min (total exposure of 352 min) gave bends in 3 out of 8 subjects. The VPM residual nitrogen table requires 409 min of decompression (total exposure of

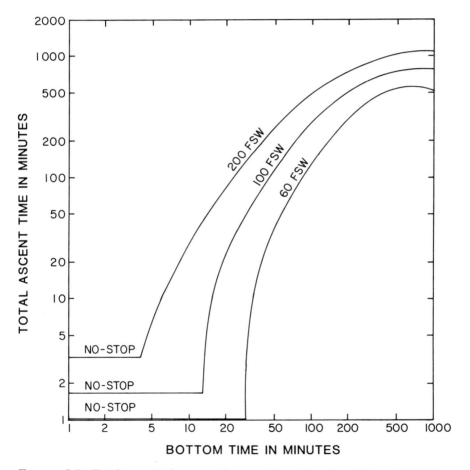

FIGURE 8.3. Total ascent times vs. bottom times for 60-, 100-, and 200-fsw exposures using a breathing mixture of 75% helium and 25% oxygen. Residual nitrogen has been taken into account. The curves begin to slope downward near the 1000-min bottom time as helium uptake falls below nitrogen elimination in the slower tissues.

529 min) and is again very conservative. The USN SCUBA tables for helium do not extend to this bottom time, but extrapolations and comparisons with other Bühlmann titrations suggest that their total decompression times in this region would be far too short (Hoffman 1985).

The total ascent times vs. bottom times are plotted in Fig. 8.3 for 60-, 100-, and 200-fsw exposures using a breathing mixture of 75% helium and 25% oxygen. An interesting feature of these curves is that they begin to slope downward near the 1000-min bottom time as helium uptake falls below nitrogen elimination in the slower tissues. This is another manifestation of the counterdiffusion phenomenon (Lambertsen and Idicula 1975).

Discussion

The immediate goal of this study was not to produce an operational set of diving tables but instead to determine whether a reasonable and comprehensive set of such tables could be computed from a bubble nucleation model using a modest number of assumptions, equations, and parameter values. The answer to this question quite obviously is yes. In fact, it is a remarkable feature of the nucleation approach—evident already in this naive formulation—that the usual proliferation of free parameters, such as M-values (U.S. Navy 1970) or tolerated overpressures (Bühlmann 1984), can be avoided.

One by-product of this investigation is an improved understanding of the practical decompression tables now in use. It is evident, for example, the profuse bubble formation is permitted by such tables, particularly during dives of short duration. Meanwhile, the number of primary bubbles, i.e., bubbles that form directly from nuclei rather than from other bubbles, is allowed to vary widely. The common assumption that the volume of released gas is critical seems still to be viable providing allowance is made for the body's ability to dissipate free gas at a useful rate.

REFERENCES

Beckman EL, Smith EM (1972). Tektite II: Medical supervision of the scientists in the sea. *Texas Reports Biol Med* 30: 155–169

Bühlmann AA (1984). *Decompression-Decompression Sickness*. Springer-Verlag, New York (Available through: Best Publishing Company, P.O. Box 1978, San Pedro, CA 90732)

Butler BD, Hills BA (1979). The lung as a filter for microbubbles. *J Appl Physiol: Respirat Environ Exercise Physiol* 47: 537–543

D'Arrigo JS (1978). Improved method for studying the surface chemistry of bubble formation. *Aviat Space Environ Med* 49: 358–361

Davson H (1964). *A textbook of General Physiology,* 3rd ed. Churchill, London, p 185

Flynn ET, Catron PW, Bayne CG (1981). *Diving Medical Officer Student Guide.* Naval Diving and Salvage Training Center, Panama City, FL

Hennessy TR, Hempleman HV (1977). An examination of the critical released gas volume concept in decompression sickness. *Proc Roy Soc Lond B* 197: 299–313

Hoffman DC (1985). On the use of a gas-cavitation model to generate prototypal air and helium decompression schedules for divers. PhD thesis. Univ. of Hawaii, Honolulu

Lambertsen CJ, Bardin H (1973). Decompression from acute and chronic exposure to high nitrogen pressure. *Aerospace Med* 44: 834–836

Lambertsen CJ, Idicula J (1975). A new gas lesion syndrome in man, induced by "isobaric gas counterdiffusion." *J Appl Physiol* 39: 434–443

Paganelli CV, Strauss RH, Yount DE (1977). Bubble formation within decompressed hen's eggs. *Aviat Space Environ Med* 48: 1429–1439

Royal Naval Physiological Laboratory (1968). *Air Diving Tables.* Her Majesty's Stationery Office, London

Schilling CW, Werts MF, Schandelmeier NR (eds) (1976). *The Underwater Handbook: A Guide to Physiology and Performance for the Engineer.* Plenum Press, New York

Strauss RH (1974). Bubble formation in gelatin: Implications for prevention of decompression sickness. *Undersea Biomed Res* 1: 169–174

Strauss RH, Kunkle TD (1974). Isobaric bubble growth: A consequence of altering atmospheric gas. *Science* 186: 443–444

U.S. Department of the Navy (1970). *U.S. Navy Diving Manual* (NAVSHIPS 0994-LP-001-9010). U.S. Government Printing Office, Washington

Yount DE (1978). Responses to the twelve assumptions presently used for calculating decompression schedules. In: Berghage TE (ed) *Decompression Theory,* the Seventeenth Undersea Medical Society Workshop. Undersea Medical Society, Bethesda, MD, pp 143–160

Yount DE (1979a). Application of a bubble formation model to decompression sickness in rats and humans. *Aviat Space Environ Med* 50: 44–50

Yount DE (1979b). Skins of varying permeability: A stabilization mechanism for gas cavitation nuclei. *J Acoust Soc Am* 65: 1429–1439

Yount DE (1981). Application of a bubble formation model to decompression sickness in fingerling salmon. *Undersea Biomed Res* 8: 199–208

Yount DE (1982). On the evolution, generation, and regeneration of gas cavitation nuclei. *J Acoust Soc Am* 71: 1473–1481

Yount DE, Gillary EW, Hoffman DC (1984). A microscopic investigation of bubble formation nuclei. *J Acoust Soc Am* 76: 1511–1521

Yount DE, Hoffman DC (1983). On the use of a cavitation model to calculate diving tables. In: Hoyt JW (ed) *Cavitation and Multiphase Flow Forum—1983.* American Society of Mechanical Engineers, New York, pp 65–68

Yount DE, Hoffman DC (1984). Decompression theory: A dynamic critical-volume hypothesis. In: Bachrach AJ, Matzen MM (eds) *Underwater Physiology VIII: Proceedings of the Eighth Symposium on Underwater PHysiology.* Undersea Medical Society, Bethesda, MD, pp 131–146

Yount DE, Hoffman DC (1986). On the use of a bubble formation model to calculate diving tables. *Aviat Space Environ Med* 57: 149–156

Yount DE, Lally DA (1980). On the use of oxygen to facilitate decompression. *Aviat Space Environ Med* 51: 544–550

Yount DE, Strauss RH (1976). Bubble formation in gelatin: A model for decompression sickness. *J Appl Phys* 47: 5081–5089

Yount DE, Yeung CM (1981). Bubble formation in supersaturated gelatin: A further investigation of gas cavitation nuclei. *J Acoust Soc Am* 69: 702–708

Yount DE, Yeung CM, and Ingle FW (1979). Determination of the radii of gas cavitation nuclei by filtering gelatin. *J Acoust Soc Am* 65: 1440–1450

9

Arterial Oxygen Tensions and Hemoglobin Concentrations of the Free Diving Antarctic Weddell Seal

Warren M. Zapol, Roger D. Hill, Jesper Qvist, Konrad Falke, Robert C. Schneider, Graham C. Liggins, and Peter W. Hochachka

Diving physiology has interested scientists for over a century (Blix and Folkow 1983). The mammals and birds which dive to great depths for long periods to exploit food sources deep in the ocean have developed remarkable evolutionary adaptations to optimize their diving ability. Some of the respiratory accommodations are obvious to casual inspection, such as a small-lung-volume-to-body-size ratio, thoracic cage mobility, and circular bronchial cartilages (Kooyman 1981). Some respiratory and circulatory adaptations have been observed in the laboratory; captive seals have been forced to dive while monitored by invasive instrumentation (Swan Ganz catheters, left ventricular catheters) (Zapol et al. 1979) or have been subjected to the pressure of depth (hyperbaric chamber) (Kooyman et al. 1972). However, it has been clear for over 10 years that laboratory diving forces an abnormally profound diving reflex, including intense bradycardia and marked regional arterial vasoconstriction (Blix and Folkow 1983; Zapol et al. 1979). This intense bradycardia is far slower than that recorded in free-swimming seals with an electrocardiogram (ECG) and breakoff leads (Kooyman and Campbell 1972).

In 1976 and 1977 our research group at McMurdo Station (168°E, 77°S) studied the Antarctic Weddell seal (*Leptonychotes weddelli*), a champion pinniped diver, capable of diving over 1 hr to depths of over 500 m (Kooyman 1981). Our studies were performed in the Eklund biological laboratory, and we examined organ biochemistry (Hochachka et al. 1977), regional blood flow with microspheres (Zapol et al. 1979), and fetal physiology (Liggins et al. 1980). Often during those laboratory studies we considered the importance of repeating the experiments in the field, realizing that many measurements during forced diving might not reflect the values in freely exercising, free-swimming seals, but we were sobered by the difficulties of monitoring heart rate and velocity, and of sampling the arterial blood of a free-swimming Weddell seal. There were three obvious problems: the pressure was high enough to cause seawater to leak into instrumentation (the seawater pressure at 500 m is 750 psi), the temperature was low enough to cause blood samples to freeze (McMurdo

Sound has a nearly constant temperature of $-1.9°C$), and there was no way to reliably retrieve blood samples or data records from a free-swimming seal.

Each member of our team contributed toward solving these problems. The central developments were the design, construction, and field testing of both a submersible 64K RAM microprocessor monitor and a blood sampling system designed by Dr. Roger Hill (1986). Also important was the development of a safe anesthetic technique for arterial cannulation and a simple, percutaneous procedure for implanting a fetal ECG electrode requiring only local anesthesia. In 1978, Dr. Gerald Kooyman (1981) demonstrated for us a captive diving technique at McMurdo which was vital to our field research and allowed us to reliably recover both monitoring data and blood samples. This chapter will summarize a small segment of those results on diving oxygen exchange which have been reported in greater detail elsewhere (Qvist et al. 1986).

Techniques

Adult male seals weighing 300–400 kg were selected from seal colonies on the annual ice near the shore of Ross Island, Antarctica, and sledged to a site at which two holes of 1-m diameter had been drilled through the 3-m-thick annual sea ice of McMurdo Sound. The field site was chosen to be sufficiently distant from any natural ice cracks that an instrumented seal released at this site would be obliged to return to the drilled holes to breathe. During general anesthesia (consisting of intramuscular 0.3 mg/kg Ketamine for induction, followed by mask inhalation via a to-and-fro circuit of 1–4% halothane in oxygen), a small skin incision over a fore-flipper artery allowed us to place an aortic catheter (3.2-mm O.D.) via an arteriotomy. Following catheterization, ECG leads were placed subcutaneously in the equivalent of "right arm" and "left leg" positions. In the 16 to 24 hr allowed for recovery from anesthesia, two neoprene patches were glued to the seal's dorsal fur with a fast-setting cyanoacrylate adhesive (Loctite 422). A microcomputer monitor and blood sampler, both described in detail elsewhere (Hill 1986), were bolted to these patches and connected to the ECG electrodes and aortic catheter, respectively.

The monitor sampled and stored records of heart rate, depth, swimming activity, and aortic blood temperature, as well as controlled the blood sampling roller pump. Either a single or a set of sequential blood samples could be initiated at any part of the dive by appropriate programming of the microcomputer monitor.

After full recovery from general anesthesia, the instrumented seal was released into the water through one of the drilled holes, which was subsequently blocked. A portable fish hut, with a hole in its floor, was placed over the other hole to act as a field laboratory. The seal was thus forced to surface inside the field laboratory between dives to breathe. The laboratory contained a small computer (Zenith Z-90 with 192 kbytes of memory and two 640 kbyte floppy disk drives) to which physiological and environmental data were transmitted through fiber-optic cables during a brief connection when the seal surfaced. At this time the instructions for obtaining the next blood sample were transmitted from the laboratory computer to the microcomputer monitor.

At the conclusion of an experiment the second hole was reopened, and when the seal hauled out onto the ice, the aortic catheter and ECG leads were removed and the monitor and blood sampler were removed. Finally the seal was sledged back to its original capture site and released.

FLUSH SOLUTION

Clotting was inhibited by flushing all the blood collection lines (and the aortic catheter) with a solution containing heparin (100 units/ml), and bonding heparin to the inside surfaces of the sample bags and collection lines (TDMAC) (Leininger et al. 1966). Freezing was a problem because the normal saline flush solution froze after several hours immersion in seawater. Blood samples, however, did not freeze for up to an hour. A concentrated (5 M) sodium chloride solution was added to the normal saline in the flush bag to raise its salinity to 0.6 M. Since contamination of blood samples with the high-molarity flush solution rendered them useless for P_{O_2}, P_{CO_2}, hematocrit, and pH analysis, it was essential to detect inadvertent dilution with flush solution. Hematocrit measurements provided the most useful field indicator of dilution, as the high osmolarity of flush solution caused dehydration of the red blood cells and a concomitant increase of the ratio of hemoglobin to hematocrit. The plasma sodium concentration of the blood samples was measured after returning from Antarctica, and those samples with an elevated sodium concentration were not included in the gas tension analysis.

BLOOD SAMPLING SYSTEM

Arterial blood was sampled at predetermined depths during descent, during ascent, or after certain dive durations had elapsed. Blood was collected with two different techniques (Hill 1986), each with its own advantages. One technique provided a 50-ml blood sample which was used for biochemical sampling and arterial P_{N_2} measurements which are reported elsewhere (Falke et al. 1985). Twenty-two single arterial samples were obtained using this technique. A second blood-collecting system provided up to eight sequential blood samples during a single dive. This system, described by Hill (1986), collected blood sequentially into a series of nine 10-ml syringes. The first syringe was filled with flush solution from the sample line and was discarded. The optimum pumping speed for uninterrupted serial sampling was found to be 20 ml/min, or 30 sec for each 10-ml syringe, and produced eight sequential samples. When sampling was discontinuous, for example, with a 5-min collection interval, two syringes were filled every 5 min. Blood remaining in the sample line filled the first of the two syringes and was discarded. This blood-collection protocol reduced the number of useful samples but provided four undiluted blood samples drawn at flexible intervals during a single dive. With this technique 31 blood samples were drawn during 9 free dives of up to 37-min duration.

An insulated field hut was erected on the sea ice near the diving hut to provide a stable thermal environment. Pa_{O_2}, Pa_{CO_2}, and pHa were measured and recorded at 37°C with an automated self-calibrating blood-gas laboratory (ABL-30, Radiometer, Copenhagen); barometric pressure was frequently recorded. As aortic temperatures during diving were near 37°C, we report all blood-gas tensions at this temperature for consistency. Hemoglobin measurements were made with another automated and frequently calibrated device (OSM-2, Radiometer). All data presented are mean $\pm$ SD.

HEMOGLOBIN AND HEMATOCRIT

Dives were partitioned into short duration ($<$ 17 min, probably feeding dives) and longer periods ($>$ 17 min, probably exploratory dives), based upon finding a normal postdive arterial pH (pHa) and arterial base excess (BEa) value in all dives of less than 17-min duration, whereas all dives of 17-min duration or greater had significant decreases of pHa and BEa and an increased blood lactate in the early surface recovery period.

The 53 arterial hemoglobin concentration measurements obtained in 32 dives are displayed in Fig. 9.1. During diving the arterial hemoglobin concentration (Hb) increased rapidly from a resting value of 15.1 $\pm$ 1.1 g% (mean $\pm$ SD). The rate of hemoglobin increase was approximately 0.85 g% $\times$ min^{-1} during the first 10 min of diving in all dives. This remarkably consistent increase in arterial hemoglobin concentration was accompanied by an increase of the arterial hematocrit (Hct), from 38.3 $\pm$ 2.85% to approximately 55%. After 10–15 min of diving there was little further increase of Hb. The highest values of Hb and Hct were sampled late during long dives: After a dive of 26.3 min the Hb and Hct were 25.4 g% and 59.3%, respectively. Both values were significantly higher (P $<$ 0.01) than the values obtained late in dives lasting less than 17 min. The Hb and Hct returned to their resting values within 12–16 min after surfacing if the seal did not immediately undertake a further long drive.

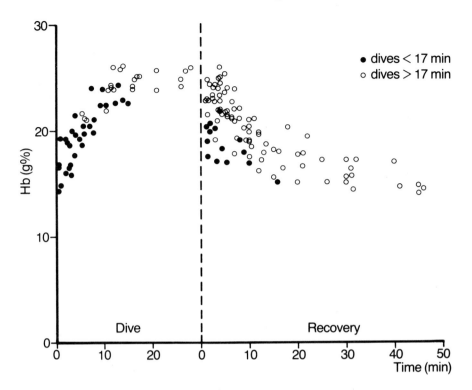

FIGURE 9.1. Arterial hemoglobin changes during diving and after resurfacing. Dives were divided into short (< 17 min) and long (> 17 min) dives. Serial blood sampling during long dives showed that Hb concentration stabilized within 10–12 min. The rate of rise of Hb was close to 0.85 g%/min during the first 10 min. The rate of decrease during recovery was similar.

ARTERIAL OXYGEN TENSIONS

The 55 arterial Po_2 (Pa_{O_2}) measurements we obtained from 33 dives are presented in Fig. 9.2. The surface Pa_{O_2} ranged from 56 to 100 mmHg (78.1 ± 12.9 mmHg), with the highest values recorded during brief resting periods between sequential dives. Blood samples that were collected within the first minute of descent gave oxygen tensions as high as 232 mmHg, well above the atmospheric Po_2 of 130–140 mmHg (atmospheric pressure was 718 to 740 mmHg), evidencing compression hyperoxia. In seven short dives (8- to 17-min duration) the Pa_{O_2} of blood sampled up to 2 min before surfacing was as low as 20.4 mmHg (mean 24.5 ± 2.9 mmHg). During these short dives to depths of 200–300 m, the heart rate (HR) at sampling time was high (average 40 bpm, range 30–65). The Pa_{O_2} value we measured was 18.2 mmHg (28% saturation) in a blood sample obtained 1 min before the conclusion of a 27-min dive.

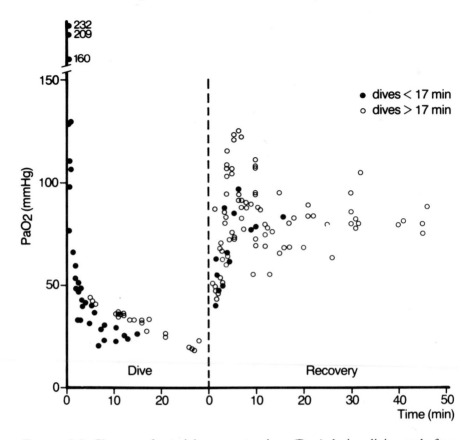

FIGURE 9.2. Changes of arterial oxygen tensions (Pa_{O_2}) during diving and after resurfacing. Early diving compression hyperoxia is apparent. The lowest Pa_{O_2} recorded was 18.2 mmHg at the end of a 27-min dive. Similar low Pa_{O_2} values were recorded at the end of short dives, i.e., dives that ended before 17 min. The highest postdive Pa_{O_2} values were recorded after dives of long duration.

Review

As we have shown that arterial Hb values increase by nearly 60% during the first 10–12 min of both long and short dives (Fig. 9.1), we now suggest that the source of this influx of oxygenated red blood cells (RBC) is the spleen. Table 1 provides the spleen weight as a percentage of body weight for selected terrestrial and marine mammals. The Weddell seal has the highest spleen weight as a percentage of body weight of any reported mammal including the southern elephant seal, which is also believed to be a prolonged diving seal (Bryden 1971). In contrast, baleen whales and porpoises have small spleens, constituting only 0.02% of body weight (Zwillenberg 1959, 1985).

TABLE 9.1. Splenic scaling for terrestrial and marine mammals.

Species	Body weight (kg)	Spleen autopsy weight (g)	Spleen autopsy weight % body weight	In vivo spleen weight % body weight	RBC storage capacity %
Weddell seal $n=13$	440 ± 70	3900 ± 2000	0.89 ± 0.214	7	60*
Harbor seal $n=12$	24 ± 19.2	—	0.40 ± 0.152	—	—
Horse $n=15$	554 ± 37	—	0.3*	1.9	54
Sheep $n=20$	45 ± 4.5	91 ± 12.5	0.20 ± 0.031	1.2	26
Dog	—	—	0.22	1.9*	20
Man	70	150 – 200	0.25 – 0.29	—	<10

* Author's estimate.

In terrestrial species, such as the sheep (Turner and Hodgetts 1959) and the horse (Persson et al. 1973), the spleen has been shown to serve as a dynamic RBC reservoir containing, respectively, 26% and 54% of the total RBC mass during rest. Excitement, exercise, or the injection of epinephrine caused the ovine spleen to contract and increased the hematocrit by 25% or more (Turner and Hodgetts 1959). There was no indication that other organs or tissues provide a reversible red blood cell storing capacity after splenectomy. The pathologically enlarged human spleen contracts markedly after epinephrine injection, producing a leuko-cytosis (Schaffner et al. 1985), but there is little increase of either the circulating hemoglobin or hematocrit (Schaffner, personal commu-nication to Zapol).

In seals the diving reflex is characterized by profound sympathetic vasoconstriction of the peripheral vasculature (Blix and Folkow 1983). We suggest that during diving the Weddell seal capitalizes upon this sympathetic response to induce constriction of its very large spleen, thus injecting large quantities of oxygenated red cells into the central circu-lation. Splenic smooth muscle might also be stimulated to contract as a result of low arterial oxygen tensions or of increased circulating catechol-amine levels. Since the spleen is an active storage vessel, determinations of its weight at autopsy may be misleading; e.g., the spleen weight of the sheep at autopsy is less than a fifth of the in vivo weight obtained by clamping the vascular pedicle during general anesthesia (Turner and Hodgetts 1959). Thus, the autopsy weight of this organ markedly underes-timates its red cell storage capacity in vivo. A splenic red cell storage capacity of 54% of the total red cell volume has been recorded in the horse (Persson et al. 1973).

We estimate that more than 60% of the red cell mass is stored in the resting Weddell seal spleen. We base our estimate upon two assumptions: (1) the circulating blood volume of the Weddell seal during rest is approximately 10% of total body weight, although laboratory dilution studies during stress have measured values of 15% of body weight or more; and (2) the seal's plasma volume remains constant during diving. This is partly justified because the horse's plasma volume decreased less than 4% after adrenalin injection, despite complete contraction of the spleen, while the measured circulating blood volume increased from 36 to 48 l (Persson et al. 1973). In a 350-kg Weddell seal with a resting hematocrit of 38% and a diving hematocrit of 60%, the circulating blood volume during diving could increase (due to red cell injection) from 35 to 55 l. This corresponds to an increase of the circulating blood volume from 10 to 15.7% of the body weight. The latter value is close to the values measured in seals having a high hematocrit during laboratory blood volume investigations (Kooyman 1981; Lenfant et al. 1969). This hypoth-esis is shown diagramatically in Fig. 9.3. Although measurements of in vivo spleen weight have not been reported for any seal, we can estimate

Dynamic spleen function of Weddell seals

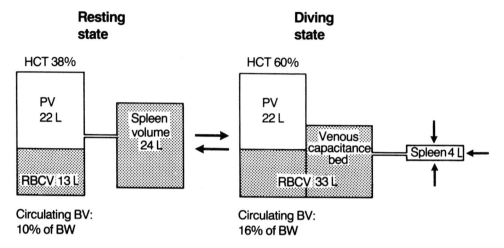

FIGURE 9.3. One hypothesis explaining the events leading to the marked increase of circulating arterial hemoglobin concentration during diving by the Weddell seal. Red blood cells (RBC) stored in the spleen at near resting O_2 and CO_2 tensions are released into the portal circulation and the subdiaphragmatic capacitance veins and then enter the central circulation via the inferior vena caval sphincter. The effects of RBC release are to maintain or increase the aortic arterial oxygen content until the reservoir is depleted (usually the first 10–12 min of long dives), reduce the buildup of CO_2 until the reservoir is emptied (a dilution effect), and reduce the initially high nitrogen tensions (a dilution effect). The splenic storage capacity of red cells may amount to 60% or more of the RBC mass, and the spleen's weight during rest probably totals at least 7% of the body weight (BW).

the in vivo resting spleen weight of the Weddell seal weighing 350 kg at 24 kg, or 7% of body weight. This estimate is conservative because Hct values of 65% have been measured and we have assumed that the splenic venous effluent has a Hct of 100%, which is unlikely. After spleen contraction, the increased circulating blood volume (BV) could be accommodated in the subdiaphragmatic venous capacitance system described in the Weddell seal and other species of seals (Blix and Folkow 1983). During diving with splenic emptying, oxygenated splenic red blood cells would transit the portal circulation, liver, and large venous cavities before entering the central circulation. Blix and Folkow (1983) reported that inferior vena cava Po_2 was higher than aortic Po_2 during laboratory studies of an unnamed seal species. Splenic contraction may have elevated the caval Po_2.

During diving the injection of large amounts of red blood cells into the central circulation with near atmospheric tensions of nitrogen may explain why arterial nitrogen tensions decrease from 2400 to 1600 mmHg (Falke et al. 1985). If a large proportion of the total red cell mass is stored in the spleen, it would not be exposed to a high alveolar P_{N_2} during descent. During the dive, diluting circulating blood with stored red cells having a near atmospheric P_{N_2} would reduce the Pa_{N_2} considerably.

Scientists studying seal physiology must strive to take measurements in free-swimming and unstressed seals. The remarkable increase of arterial hemoglobin concentration we recorded with each dive was not observed in laboratory diving. Blood sampling of freely diving Weddell seals suggests they remain aerobic and do not increase plasma lactate concentration for diving periods of up to 17 min (Guppy et al. 1986; Kooyman et al. 1983). In contrast, a laboratory dive as short as 5 min is followed by a recovery lactate washout into circulatory blood. The difference may well rest upon selective perfusion of exercising muscles in unstressed short free dives. Perfusing exercising muscles would account for the rapid reduction of aortic Pa_{O_2} we measured in short dives (Fig. 9.2). A complete explanation of this metabolic puzzle awaits the continuous measurement of regional blood flow during free diving, just as the precise anatomical identification of the source of the diving increase of arterial hemoglobin concentration, presumably the spleen, awaits future microprocessor monitoring studies upon free-swimming seals.

Acknowledgements. This work was supported by the National Science Foundation, Division of Polar Programs Grant No. 81000212, and grants from the Danish Medical Research Council and the NATO Science Fellowship Scheme. The work required field support from the National Science Foundation, ITT Antarctic Services, and U.S. Naval Support Force Antarctica. The studies were performed in accordance with Permit No. 394 granted by the National Marine Fisheries Service, U.S. Department of Commerce, NOAA.

REFERENCES

Blix AS, Folkow B (1983). Cardiovascular adjustments to diving in mammals and birds. In: Shepherd J, Abboud F (eds) *Handbook of Physiology—The Cardiovascular System III*. Williams & Wilkins, Baltimore, chap 25, p 917

Bryden MS (1971). Size and growth of viscera in the southern elephant seal, *Mirounga leonina. Aust J Zool* 19: 103–120

Falke KJ, Hill RD, Qvist J, Schneider RC, Guppy M, Liggins GC, Hochachka PW, Elliot RE, Zapol WM (1985). Seal lungs collapse during free diving: Evidence from arterial nitrogen tensions. *Science* 229: 556–558

Guppy M, Hill RD, Schneider RC, Qvist J, Liggins GC, Zapol WM, Hochachka PW (1986). Microcomputer-assisted metabolic studies of voluntary diving of Weddell seals *Am J Physiol* 250 (19): R175–R187

Hill RD (1986). Microcomputer monitor and blood sampler for freely-diving seals. *J Appl Physiol* 61: 1570–1576

Hochachka PW, Liggins GC, Qvist J, Schneider RC, Snider MT, Wonders TR, Zapol WM (1977). Pulmonary metabolism during diving: Conditioning blood for the brain. *Science* 198: 831–834

Kooyman GL (1981). *Weddell Seal: Consummate Diver.* Cambridge University Press, Cambridge

Kooyman GL, Campbell WB (1972). Heart rates in freely diving Weddell seals, *Leptonychotes weddelli. Comp Biochem Physiol* 43A 31–36

Kooyman GL, Castellini MA, Davis RW, Maue RA (1983). Aerobic diving limits of immature Weddell seals. *J Comp Physiol* 151: 171–174

Kooyman GL, Schroeder JP, Denison DM, Hammond DD, Wright JJ, Bergman WP (1972). Blood nitrogen tensions of seals during simulated deep dives. *Am J Physiol* 223: 1016–1020

Leininger RI, Epstein MM, Falb RD, Grade GA (1966). Preparation of non-thrombogenic plastic surfaces. *Trans Am Soc Artif Int Organs* 12: 151

Lenfant C, Elsner RE, Kooyman GL, Drabek CM (1969). Respiratory function of blood of the adult and fetal Weddell seal, *Leptonychotes weddelli. Am J Physiol* 216: 1595–1597

Liggins GC, Qvist J, Hochachka PW, Murphy BJ, Crease RK, Schneider RC, Snider MT, Zapol WM (1980). Fetal cardiovascular and metabolic responses to simulated diving in the Weddell seal. *J Appl Physiol* 49: 424–430

Persson SGB, Ekman L, Lydin G, Tufvesson G (1973). Circulatory effects of splenectomy in the horse I–IV: II. Effect on plasma volume and total and circulating red-cell volume. *Abl Vet Med A* 20: 456–468

Qvist J, Hill RD, Schneider RC, Falke KJ, Liggins GC, Guppy M, Elliot RL, Hochachka PW, Zapol WM (1986). Hemoglobin concentrations and blood gas tensions of free-diving Weddell seals. *J Appl Physiol* 61: 1560–1569

Schaffner A, Augustiny MD, Rainer CO, Fehr J (1985). The hypersplenic spleen, a contractile reservoir of granulocytes and platelets. *Arch Intern Med* 145: 651–654

Turner AW, Hodgetts UE (1959). The dynamic red cell storage function of the spleen in sheep. Relationship to fluctuations of jugular haematocrit. *Aust J Expl Biol* 37: 339–420

Zapol WM, Liggins GC, Schneider RC, Qvist J, Snider MT, Creasy RK, Hochachka PW (1979). Regional blood flow during simulated diving in the conscious Weddell seal. *J Appl Physiol* 47 (5): 968–973

Zwillenberg HHL (1959). Uber die Milz des Braunfisches. *Zeitschr Anat Entw* 121: 9–18

Zwillenberg HHL (1985). Die mikroskopische Anatomie der Milz der Furchen-wale. *Acta Anat* 32: 24–39

Part III Physiology of Exposure to Altered *G*-Force

10

Human Physiological Limitations to G in High-Peformance Aircraft

R.R. Burton

Introduction

The accelerative force (principally $+G_z$)[1] of the environment of the aerial combat maneuver (ACM) of high-performance aircraft has two dimensions that determine human tolerances: (*a*) G level and (*b*) G duration. Typically, G-level tolerances are measured using light-loss criteria, such as grayout (loss of peripheral vision), blackout (loss of central vision), or G-induced loss of consciousness (G-LOC), whereas G-duration physiological limitation is a function of pilot fatigue. A review of G tolerance determinations is available (Burton et al. 1974).

In order for pilots to maintain vision and consciousness while they are above 5 G (even while wearing anti-G suits), they must perform an anti-G straining maneuver (Burton 1986). The anti-G straining maneuver (AGSM) is a voluntary physical tensing of the muscles of the arms and legs along with a forced respiratory exhalation against a closed or partially closed glottis. This effort is continued for approximately 3 sec, which is then followed by a relaxation of the tensed muscles with a rapid completion of exhalation and an abrupt inhalation which requires less than 1 sec. This maneuver when executed properly in a continuous cyclic fashion results in an increase in $+G_z$ light-loss mean tolerance of as much as 4 G (Burton 1986). This AGSM, frequently referred to as the M-1 maneuver, was developed by Dr. Earl Wood and associates during World War II (Wood et al. 1946) and remains today, in conjunction with the anti-G suit, as the principal anti-G method used by pilots flying high-performance fighter aircraft. The cardiovascular and pulmonary physiology of the AGSM is well known (Burton et al. 1974; Wood et al. 1981).

Although the AGSM is extremely effective in increasing G tolerance to levels required for flying fighter aircraft (up to $+9 G_z$), it takes considerable physical effort and pilots rapidly become fatigued. This fatigue factor due to sustained G, although an important determinant in characterizing G tolerance, had not received any attention in the laboratory. Now,

---

[1] Inertial accelerative forces are designated as G. The + identifies force directed toward the feet, and the subscript z identifies force along the longitudinal axis of the body.

however, because of an emerging importance in USAF operations, laboratory research has begun to focus on *G*-induced fatigue in an effort to reduce its adverse impact on ACM tolerance. A review of this research follows.

Fatigue-Limiting *G*-Tolerances

The ACM is an environment of multiple *G* levels (2 to 9 *G*) which continues for an indefinite period of time (few seconds to several minutes). Because of this extreme variability and therefore difficulty, a method to measure ACM tolerance has only recently been devised for laboratory use on a centrifuge (Burton and Shaffstall 1980).

ACM TOLERANCES

Tolerance to the ACM with fatigue as an end point is determined in the laboratory using a variable-*G* profile called a simulated aerial combat maneuver (SACM), shown in Fig. 10.1 (Burton and Shaffstall 1980). Each *G* level of the SACM is 15-sec long, and the profile continues until the subject becomes too fatigued to continue. The duration of time (sec) until the fatiue end point is reached is that subject's SACM *G* tolerance. Several modifications of this approach have been developed, but the principles of this test remain the same. This profile has been validated as

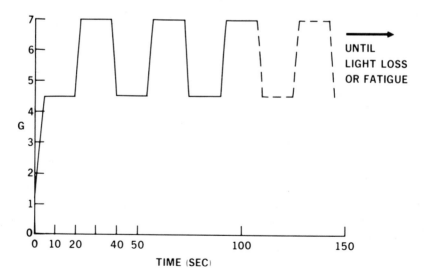

FIGURE 10.1. SACM profile used on the USAFSAM centrifuge to determine fatigue tolerances of experimental subjects (Burton and Shaffstall 1980).

TABLE 10.1. SACM tolerance times (control values only) are compared from several studies with subjects seated in an upright (13–15°) seat, using the standard 5-bladder anti-*G* suit (unless indicated) and performing the AGSM as required.

SACM	*N*	Time (sec) ($\overline{X} \pm$ SE)	Study objective and reference
4.5/7 *G*	6	102 ± 11	Evaluate positive pressure breathing (PPB) (Shaffstall and Burton 1979)
4.5/7 *G*	7	161 ± 25 170 + 17	Develop and validate the SACM profile (Burton and Shaffstall, 1980)
4.5/7 *G*	9 * 8 * 7 *	195 ± 34 180 ± 31 232 ± 33	Evaluate physical conditioning SACM tolerance (Epperson et al. 1982).
4.5/7 *G*	11 †	245 ± 33	See Epperson et al. (1982) above (Tesch et al. 1983)
4.5/7 *G*	5	213 ± 41	Evaluate anti-*G* suit designs (Shaffstall and Burton 1980)
4.5/7 *G*	9 11	222 ± 39 202 ± 34	See Epperson et al. (1982) above (Spence et al. 1981)
5/9 *G* §	7	74 ± 7	Evaluate PPB (Burns and Balldin 1983).
3.5/5.5 *G*	10 †	335 ± 52‡	See Epperson et al. (1982) above (Balldin et al. 1985)

* Separate experimental groups before initiating physical conditioning longitudinal study—all were SACM trained subjects.
† All subjects were male fighter pilots from the Royal Swedish Air Force.
‡ Without anti-*G* suit.
§ 10-sec duration at each level of *G*.

reproducible within a subject population and has been used in several research studies in our laboratory (Burns and Balldin 1983; Burton and Shaffstall 1980; Epperson et al. 1982; Shaffstall and Burton 1979, 1980; Spence et al. 1981) and in the Karolinska Institute, Stockholm, Sweden (Balldin et al. 1985; Tesch et al. 1983; Tesch and Balldin 1984), with considerable success as regarding reproducibility between laboratories, use in physiological studies, and evaluation of anti-*G* methods (Table 10.1).

The relationship of fatigue tolerance to *G*-level tolerance in subjects has been determined using reclined seats to increase their *G*-level tolerance and is shown in Fig. 10.2 (Burton and Shaffstall 1980). As relaxed light-loss tolerance is increased and less AGSM is required at 7 *G,* so tolerance to fatigue is increased but at a complex exponential rate. The physiological basis for this complex relationship between fatigue and the cardiovascular requirements (*G* level) of the ACM is not completely known at this time. However, a predictive model has been recently proposed that relates fatigue to *G*-level tolerances. It has been validated with published data and appears to be reasonably accurate (Burton 1986).

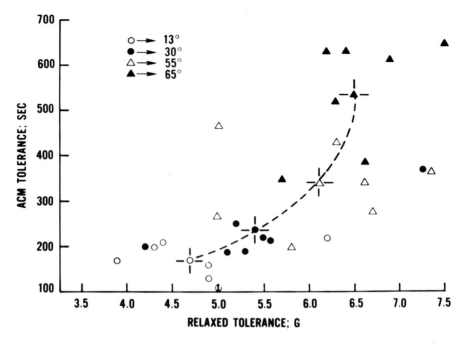

FIGURE 10.2. Relaxed (light-loss) *G* tolerances are compared with straining fatigue tolerances. Relaxed tolerances were altered by reclining subjects to various seat-back angles (Burton and Shaffstall 1980).

SUSTAINED G TOLERANCES

Since the SACM is composed of repeated short durations of sustained *G* levels, fatigue tolerances of different sustained *G* levels do relate to the SACM and therefore are compared in Fig. 10.3. Miller et al. (1959) found that fatigue prevented subjects from continuing sustained low *G*-level exposures ranging from 3 to 5 *G* with and without anti-*G* suits. Although they limited several of the centrifuge exposure durations with arbitrary fixed values, several subjects did become fatigued before reaching these arbitrary limits. They therefore established that even at low *G* levels beginning at 3.5 *G*, where neither the anti-*G* suit nor the AGSM is required, fatigue can become a limiting factor.

The 4.5-*G* sustained exposure has recently been repeated in our laboratory, allowing six subjects,[2] all wearing anti-*G* suits, to continue until fatigue intervened. We found the body-fatigue end point not to be as clearly recognizable for all subjects as found for an SACM to fatigue. Three subjects got uncomfortably tired, while the others stopped the run

---

[2] The voluntary informed consent of the subjects used in this research was obtained in accordance with AFR 169-3.

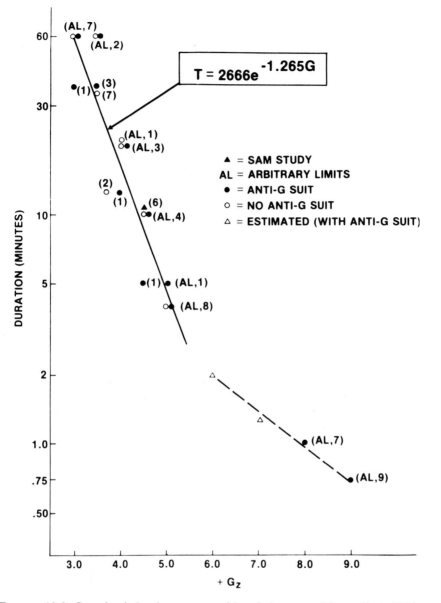

FIGURE 10.3. Sustained *G* tolerances to either fatigue or arbitrary limit (AL)—duration established before the study. Number of subjects per data point is in parentheses. Data for 3 to 5 *G* are from Miller et al. (1959)—the equation was developed by this reviewer. Data for 6 *G* and above are from in-house studies (Burton 1986; Burton et al. 1974) and estimates derived from research experience (see text).

because of specific neck muscle fatigue, abdominal muscle fatigue, or discomfort from the inflated anti-*G* suit. Our group "fatigue-tolerance" mean ± SE for all six subjects at 4.5 *G* was 623 ± 68.7 sec, which is in reasonable agreement with the results of Miller et al. (1959) (Fig. 10.3, SAM study).

With the use of their data and considering all of their published values, including the arbitrary limits (AL) values, an inverse exponential relationship relative to *G* level was calculated:

$$T = 2666e^{-1.265G} \tag{10.1}$$

in which *T* = tolerance to fatigue, or AL (min)

      G = *G* level (3- to 5-*G* limits)

At higher *G* levels, above 5 *G*, where the AGSM is required, fatigue tolerances as related to *G* levels are not continuous with Eq (1) (Fig. 10.3). Interestingly, fatigue tolerances at these higher *G* levels are greater than predicted with Eq. (10.1), even though considerably greater physical effort is required because of the AGSM.

These fatigue tolerances for 8 and 9 *G*, as was the case for the Miller study, are data from studies using "arbitrary" duration limits—established before the study began—which approximately 50% of our subjects obtained (Burton et al. 1974). Although the 60 sec at 8 *G* and 45 sec at 9 *G* were limits "arbitrarily" set, most of the subjects that made the limits were extremely fatigued, so that the absolute (unlimited) fatigue values would not have been much greater. Also, it is important to realize that these subjects were well trained and very proficient in doing the AGSM. In addition, some selection had occurred, since the remaining half of these subjects did less well; e.g., in the 8-*G* study, 5 of the total 12 subjects had a mean tolerance of 32 sec, which was their fatigue end point.

The fatigue tolerances of 75 sec at 7 *G* and 120 sec at 6 *G*, shown in Fig. 10.3, are estimates only, since subjects have never been allowed to continue at these *G* levels to fatigue. These estimates, however, were not completely arbitrarily made but were based on our extensive centrifuge experience (Burton 1986; Burton and Krutz 1975; Burton et al. 1974). Specifically, we do know that all subjects with the anti-*G* suit routinely tolerate 60-sec exposures to 7 *G* without reaching a fatigue end point and do the same at 6 *G* even without an inflated anti-*G* suit. Unlike the lower *G* levels, where the anti-*G* suit apparently is not useful in reducing fatigue, at 6 *G* and above the suit is very effective in extending fatigue tolerance since it increased relaxed *G* tolerances by approximately 1 *G*, thereby reducing the amount of AGSM required to tolerate *G*.

Since the relationship of the duration (fatigue) tolerance to the *G* level becomes discontinuous between 5 and 6 *G*—which incidentally is the *G* level when the AGSM begins to be required—it appears that different energetics are involved; i.e., the "relaxed" *G*-tolerance fatigue-related energetics are no longer limiting above 5 *G*.

Subjective Effort and Fatigue in SACM Tolerances

Effort required to tolerate a 3.5/5-G SACM (similar to the one shown in Fig. 10.1) to fatigue without the subjects wearing an anti-G suit was subjectively measured on the Borg Perceived Effort Score (Borg 1982) at 8.5 ± 1.5 (mean ± SD) to 9.2 ± 1.2 for two groups of 10 subjects each (Balldin et al. 1985). This effort-rating scale is logarithmic, with a minimum score of 0.5 (very, very weak) to a maximum effort score of "slightly" over 10. Scores of 8 and 9 as found for the SACM are situated between very strong, which is a 7, and very, very strong (almost max), which is 10. Obviously, tolerating a SACM to fatigue is perceived to require extreme effort.

Subjective fatigue evaluations have been determined immediately after SACM exposures (Burton 1980; Epperson et al. 1982). Fatigue was scored from 20 to 0, with the lowest score indicating maximum fatigue. Pre-SACM fatigue scores ranged from means of 12.4 to 16.7 in four study groups with immediate post-SACM scores ranging from 6.2 to 8.8. These fatigue scores indicate that subjects are tired, but not extremely fatigued.

Although the fatigue end point at 4.5-G sustained exposure (discussed earlier in this chapter) was not perceived by the subjects to be as clearly identifiable as was fatigue at the end of the 4.5/7-G SACM, the same subjects exposed to both environments to "fatigue" had similar mean fatigue scores, and, interestingly, recovery for the 4.5-G sustained group was slightly slower—72% recovery after 20 min of 1-G rest compared with 88% recovery after the SACM (unpublished data). Fatigue therefore develops subjectively similarly in response to various acceleration environments above 3 G and is perceived to be a limiting tolerance factor.

Physical Conditioning Effects on SACM Tolerance

The effects on changes in physical conditioning on SACM fatigue tolerance were first examined by Epperson et al. (1982, 1985), in a longitudinal study late in the 1970s. They weight-trained seven subjects and aerobically trained eight others over a period of 12 weeks. Both groups significantly improved their appropriate physical condition, but only members of the weight-trained group significantly increased their SACM tolerance (see Fig. 10.1 and Table 10.1) to fatigue and by 77%. Incidentally, if the increase in the controls is considered, the net increase in SACM tolerance is 53%.

Although no physiological parameters associated with these changes in physical conditioning were determined during the SACM, the fact that an increase in $\dot{V}o_{2max}$ of 7.5% (P < 0.05) in the aerobically trained group did not affect the fatigue tolerance indicates that SACM tolerance is not limited by aerobic capacity. Interestingly, Tesch and Balldin (1984) noted that the more successful G riders had lower aerobic power and capacity.

In addition, Epperson et al. (1985) found that the abdominal muscular strength increased most significantly over the other muscle groups of the body and was highly correlated with the improvement in SACM tolerance. Since this muscle group is considered extremely important in performing an effective AGSM, Spence et al. (1981) and Balldin et al. (1985) had subjects improve only this group of muscles. Neither laboratory found statistically significant increases in SACM fatigue tolerance. These findings demonstrate that improved *total body* muscular strength and related energetics are required to reduce the fatigue of the SACM.

Tesch et al. (1983) duplicated the total body weight-training portion of the study by Epperson et al. (1982) using fighter pilots and found, as expected, a significantly increased SACM tolerance (of 39%) to fatigue. Their study considered in some detail both the aerobic and anaerobic physiological and histochemical changes as related to the increases in physical condition and in the SACM tolerance.

Histochemical parameters examined included changes in the percentage of fast- and slow-twitch muscle fibers, mean fiber area, capillary density, and capillary–muscle fiber ratio. No statistically significant differences were found. These findings were expanded with an additional comparative (not longitudinal) study examining SACM tolerances in pilots and nonpilots (Tesch and Balldin 1984). They found no statistically significant correlation between muscle-fiber-type composition, fiber size, or capillary supply with SACM tolerances.

The physiological parameters examined by Tesch et al. (1983) included muscle strengths during both fast and slow contractions, muscle endurance, anaerobic and aerobic capacities, and lean body mass. The only statistically significant changes were an increase in knee extensor strength during slow contractile speeds ($P < 0.001$, an increase of 17%) and an increase in anaerobic power ($P < 0.01$, up 14%). Anaerobic power was the amount of work generated by the quadriceps muscle during a 50-sec exercise task. The researchers concluded that the increase in SACM tolerance time resulted from the need for less muscular effort for the AGSM because of both the increased muscular strength (but not associated with an increase in fast-twitch fibers) and the increased anaerobic power (capacity), which is a response to the physical training of the entire body. They further hypothesized that additional physiological benefits from the strength training that would be of value in increasing G tolerance were (a) increased sympathetic drive and (b) decreased skeletal muscle capillary density which would reduce the amount of leg blood pooling. The latter, although not statistically significant, was reduced by 25% in their study.

A paradoxical situation exists, however, regarding muscle capillary density in the legs during G. Although decreased capillary density reduces the opportunity for leg blood pooling—which is considered beneficial for increasing light-loss G tolerance—it also has the potential for reducing blood flow to contracting muscles. Obviously, reduced blood flow would contribute significantly to an increased rate of muscle fatigue (Simonson and Lind 1971). Therefore a condition that would increase *G*-level tolerance could reduce *G*-duration tolerance.

Anaerobic Capacity in SACM Tolerance

Utilization of anaerobic capacity as it relates to the SACM has been examined in three studies that exposed subjects to fatigue, using various *G* levels (Table 10.2). Different *G* levels require of the subject different levels of effort necessary to perform an adequate AGSM. The theoretical level of effort required to tolerate various *G* levels using various anti-*G* methods has been calculated in mmHg using intrathoracic pressure (Burton 1986).

The relationships of anaerobic capacity utilization, as measured with blood lactate levels, to the effort required to tolerate the *G* levels and the durations of *G* exposure are found in Table 10.2. All of these subjects were exposed to these various *G* profiles until each had reached the same subjective end point of fatigue (see earlier section in this chapter on subjective effort and fatigue). If we consider, therefore, that all subjects (as a group) were fatigued equally, a relationship between tolerances of *G* levels, including duration of *G* exposure and anaerobic capacity utilization, can be acknowledged.

Higher blood lactate levels are found in subjects exposed to higher *G*

TABLE 10.2. Peak blood lactate levels following exposures to levels of *G* to fatigue.

G profile	Duration (sec)	*N**	AGSM (mmHg)†	Blood Lactates (mg%)	References
4.5 *G* (sustained)	623	6	0	29.8	Burton et al. 1987
3.5/5.5 *G* (SACM)‡	335	10	38	37.8	Balldin et al. 1985
4.5/7 *G* (SACM)	245	6	50	47.7	Tesch et al. 1983
	338§	6	50	56.7	
4.5/7 *G* (SACM)	112	4	50	31.8	Tamir et al. 1988
7–9 *G* (sustained)	32	4	100¶	42.4	Burton et al. 1987

* *N* = number of subjects per study.
† Intrathoracic pressure required of the AGSM to tolerate maximum *G* level of the *G* profile (see Burton 1986 for calculation method).
‡ Anti-*G* suit was not used.
§ Exposure duration after strength-training program.
¶ Subjects performed maximal AGSM (usually more than required to tolerate *G* levels).

levels and somewhat independent of G duration. For instance, blood lactates of 29.8 mg% are found after exposing subjects to 623 sec of 4.5 G, whereas much higher blood lactates, 42.4 mg%, occur after only 32 sec at the higher 7–9-G sustained exposures. This ability to utilize more of the anaerobic capacity at high G appears to be the physiological basis for greater relative fatigue tolerances above 5 G as compared with lower G levels, as shown in Fig. 10.3.

If the duration of G exposures is kept relatively constant—335 sec for the Balldin study (Balldin et al. 1985) and 338 sec for the Tesch study (Tesch et al. 1983)—and G levels are varied, the higher G level is tolerated since more anaerobic capacity is utilized (higher blood lactate levels).

Tesch et al. (1983) increased the anaerobic capacity of their subjects through strength training. This increased anaerobic capacity allowed for greater utilization of this capacity with a greater SACM duration tolerance and higher blood lactate levels—47.7 mg% for the controls and 56.7 mg% for the weight-trained.

In a different manner, the direct relationship of anaerobic capacity utilization upon SACM tolerance has been determined by Tesch et al. (1983) in their strength-training study. They considered the utilization of an individual's anaerobic capacity while exposed to the same SACM to fatigue. They found, on an individual-subject basis, a significant (P < 0.01) direct correlation between G tolerance (min) to the 4.5/7-G SACM until fatigue and blood lactate levels:

$$L = 3.53 + 0.496t \qquad (10.2)$$

in which L = blood lactic acid (mmol/l)
    t = time of SACM to fatigue (min)
Simply, Eq. (10.2) shows that those subjects that tolerated the SACM for the longest duration had the highest blood level of lactate. In addition, as subjects "increased their strength," as Tesch et al. (1983) found, their "peak" blood lactate levels following longer-duration SACMs increased from 47.7 to 56.7 mg% (Table 2). Subjects that are less tolerant to this same 4.5/7-G SACM with duration limits of only 112 sec have been studied in our laboratory (Burton et al. 1987). Blood lactates taken after 1-min post-SACM exposure [time of blood withdrawal used by Tesch et al. (1983)] in four fatigued subjects were only 31.8 mg% (Tamir et al. 1988). Consequently, it is reasonable to conclude that tolerance to the SACM is directly and primarily dependent on the size of the anaerobic capacity and ability to utilize that capacity.

On the other hand, anaerobic capacity cannot be utilized maximally under extreme physical demands. Four subjects were exposed to either 7, 8, or 9 G sustained to fatigue. Each subject was asked to perform a maximal AGSM regardless of the G level. Their mean duration of high-G tolerance was only 32 sec, with a mean ± SE blood lactate level of 42.4 ±

3.2 mg% (Table 10.2). As shown in Table 10.2, higher peak blood lactate levels occur if the level of *G* is lower, allowing time for greater utilization of the anaerobic capacity and greater lactate washout from the muscle tissue.

The effect of the AGSM per se upon blood lactates was determined with five subjects performing a series of five 2-min AGSMs at 1 *G* (Burton 1980). The level of their AGSM at 1 *G* was gauged by the subjects to be the same as when tolerating (the week before) the high-*G* SACMs. After the first 2-min period of SACM, blood lactates were similar for both groups (33.3 mg% for the *G* exposure and 29.0 mg% for the 1-*G* simulation); however, after the fifth exposure, the SACM blood lactates were approximately twice those of the 1-*G* AGSM level (43.2 to 23.2 mg%). Anaerobic capacity therefore appears to be utilized approximately equally between the demands of the AGSM per se and those of the accelerative forces.

Isometric Exercise Relation to SACM Fatigue

The strength of the AGSM as discussed earlier is usually measured by determining the increase in intrathoracic pressure which causes an increase in arterial pressure at eye-heart levels on an approximate 1-to-1 basis (Burton 1986; Wood et al. 1981). Since 8- and 9-G levels require approximately 100 mmHg of intrathoracic pressure—each increase in *G* above 5 *G* is equal to approximately 25 mmHg arterial pressure (Burton 1986—a 100-mmHg AGSM (near maximal effort) is required. The relative effect, and therefore importance, of changing the amount of AGSM necessary to tolerate *G* on SACM tolerance to fatigue is shown in Fig. 10.4 (Burton 1986).

The basis for the AGSM is principally isometric muscular contractions involving many of the skeletal muscle groups of the body operating at a near maximum voluntary contraction (MVC) level if 100 mmHg intrathoracic pressure is to be achieved. If, however, the isometric contractions are maintained without interruption (intermittent relaxations), cardiac return is diminished and loss of consciousness will occur (Burton et al. 1974; Wood et al. 1981). Consequently, these isometric contractions are relaxed briefly (< 1 sec) every 2 to 3 sec, resulting in a combination of isometric and isotonic efforts—see the introduction section in this chapter. Tesch et al. (1983) described the AGSM as intermittent isometric contractions that, unlike sustained isometric contractions, pump blood back toward the heart. Consequently, isometric muscular physiology should be relevant to SACM tolerances, and this relationship is now considered.

Of particular USAF operational concern and relevance to understanding the energetics of the AGSM is that an MVC can be required within

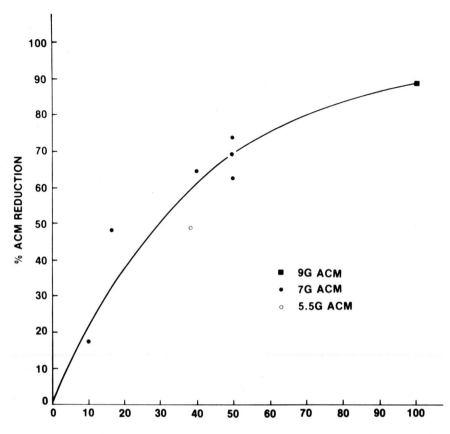

FIGURE 10.4. The reduction in SACM (ACM) tolerance time is compared with the level of AGSM in mmHg required to tolerate these *G* levels. Less AGSM was required during some 7-*G* ACM exposures because the subjects were reclined in the *G* field, thereby reducing their vertical eye-heart distance (Burton 1986).

1 sec after a pilot is in a relaxed state since operational fighter aircraft are capable of sustaining high *G* levels at onset rates of 6 to 12 *G*/sec.

At the initiation of such an abrupt vigorous static (isometric) or for that matter dynamic exercise, ATP splitting (phosphagen) provides the primary energy source for all voluntary muscular activity. This immediate phosphagen energy source is commonly called anaerobic alactate since lactic acid is not produced and oxygen is not used as the immediate energy supplier—this ATP/CP energy source is reenergized immediately at the initiation of the oxygen deficit repayment (Mole 1983).

The role of ATP splitting during the initiation of the SACM was examined using subjects exposed for 7–10 sec to an 8- or 9-*G* level that required a maximal AGSM effort (Burton et al. 1987). Blood lactates were measured immediately before and 3 min after this *G* exposure. For a group of six subjects, the lactates (mg%) of four subjects basically remained unchanged—pre-*G* lactates (mean ± SE) of 13.2 ± 3.2 and post-G lactates of 14.5 ± 3.5. However, two subjects had an increase in their blood lactate level, that is, 14.2 ± 1.9 pre-G to 24.3 ± 2.7 post-G. These data suggest that in fact the immediate source of energy for the SACM is anaerobic alactate—the response was much too rapid and intense to be aerobic, and significant levels of lactates were not produced in four subjects so that anaerobic glycolysis cannot be the energy source. However, since some increase in blood lactate did occur in two of the subjects, anaerobic glycolysis is beginning to become an important energy source at this time.

An interesting study by MacDougall et al. (1977) relates this early phosphagen energy source with biochemical skeletal muscle adaptation to heavy resistance training. Following 5 months of heavy resistance training similar to the strength-training techniques that increased SACM tolerance to fatigue (Epperson et al. 1982; Tesch et al. 1983), MacDougall et al. (1977) found significant increases in resting muscle concentrations of muscle creatine, CP, ATP, and glycogen. They postulated that an increase in muscle energy reserves would have a significant effect on length of time MVC or repeated contractions that could be sustained. They further suggested that since resting muscle glycogen concentration had increased, this type of short-duration, highly intense exercise (not unlike the AGSM) involves considerable demands on lactate-producing glycolysis in addition to phosphagens.

Since the muscular effort of the AGSM, especially at high-*G* levels, requires a high percentage of MVC, then the ability to sustain a series of high isometric contractile forces, each of 3-sec duration, is necessary. Unfortunately, the level of these contractile forces rapidly decreases as the effort is repeated (Ahlborg et al. 1972; Stull and Clarke 1971). Stull and Clarke (1971) compared the strength decay of continuous isometric and 2-sec-duration isotonic types of arm muscular contractions during a maximal effort. Since the AGSM can be considered a combination of these two types of efforts, an intermediate decay curve was estimated by using both sets of data. The levels of decay at 45 to 60 sec were approximated using this fatigue curve and were found to show respective losses of 30 and 40% of the MVC. Similar rates of loss of muscular strength of the leg quadriceps muscle have been reported (Ahlborg et al. 1972; Karlsson et al. 1975; Karlsson and Ollander 1972).

Losses of this magnitude of MVC capability would probably translate into similar decreases in the AGSM capability. A 30 to 40% decrease in the level of AGSM translates into a reduction of 1 to 2 *G* in total capability. This loss of *G* tolerance is probably the reason why subjects terminate the SACM exposure before they are exhausted—they are too fatigued to develop the level of AGSM necessary to continue the SACM yet not totally fatigued. *G*-induced loss of consciousness which occurs in pilots after tolerating an ACM for some time may be a result of this decay in the MVC. These values are consistent with those estimated using our static fatigue model (Burton 1986). Ahlborg et al. (1972) Karlsson et al. (1975), and Karlsson and Ollander (1972) reported the lowest levels of lactates in muscles fatigued at the highest MVC (50 to 98%). Blood lactate levels of only 22.5 to 45 mg% were associated with fatigued isometric muscle lactates of 100 to 195 mg% wet tissue. These are blood lactate levels similar to those found in subjects fatigued by the SACM or sustained G (Table 10.2).

Conclusion

Research data presented herein show that fatigue limits tolerance to sustained low and high G as well as the SACM. Also these data suggest that the metabolic basis of the ACM is anaerobic and that tolerance to the ACM is limited by the size of the anaerobic capacity and by the ability of the body to utilize that capacity. Consequently, research directed toward improving a pilot's tolerance to the ACM should focus on this physiological parameter.

REFERENCES

Ahlborg B, Bergstrom J, Ekelund L, Guarwieri G, Harris RC, Hultman E, Nordesjo, LO (1972). Muscle metabolism during isometric exercise performed at a constant force. *J Appl Physiol* 33: 224–228

Balldin UI, Myhre K, Tesch PA, Wilhelmsen U, Anderson HT (1985). Isometric abdominal muscle training and G tolerance. *Aviat Space Environ Med* 56: 120–124

Borg GAV (1982). Psychophysical bases of perceived exertion. *Med Sci Sports Exercise* 14: 377–381

Burns JW, Balldin UI (1983). +G$_z$ protection with assisted positive-pressure breathing (PPB). *Aerosp Med Assoc Preprints,* pp 36–37

Burton RR (1980). Human responses to repeated high G simulated aerial combat maneuvers. *Aviat Space Environ Med* 51: 1185–1192

Burton RR (1986). A conceptual model for predicting pilot group G tolerance for tactical fighter aircraft. *Aviat Space Environ Med* 57: 733–744

Burton RR, Krutz Jr RW (1975). G-tolerance and protection associated with anti-G suit concepts. *Aviat Space Environ Med* 46: 119–124

Burton RR, Leverett Jr SD, Michaelson ED (1974). Man at high-sustained +G$_z$ acceleration: A review. *Aerosp Med* 45: 1115–1136

Burton RR, Shaffstall RM (1980). Human tolerance to aerial combat maneuvers. *Aviat Space Environ Med* 51: 641–648

Burton RR, Whinnery JE, Forster EM (1987) Anaerobic energetics of the simulated aerial combat maneuver (SACM). Aviat Space Environ Med 58: 761–767

Epperson WL, Burton RR, Bernauer EM (1982). The influence of differential physical conditioning regimens on simulated aerial combat maneuvering tolerance. *Aviat Space Environ Med* 53: 1091–1097

Epperson WL, Burton RR, Bernauer EM (1985). The effectiveness of specific weight training regimens on simulated aerial combat maneuvering G tolerance. *Aviat Space Environ Med* 56: 534–539

Karlsson J, Funderburk CF, Essen B, Lind AR (1975). Constituents of human muscle in isometric fatigue. *J Appl Physiol* 38: 208–211

Karlsson J, Ollander B (1972). Muscle metabolites with exhaustive static exercise of different duration. *Acta Physiol Scand* 86: 309–314

MacDougall JD, Ward GR, Sale DG, Sutton JR (1977). Biochemical adaptation of human skeletal muscle to heavy resistance training and immobilization. *J Appl Physiol* 43: 700–703

Miller H, Riley MB, Bondurant S, Hiatt EP (1959). The duration of tolerance to positive accleration. *Aviat Med* 30: 360–366

Mole PA (1983). Exercise metabolism. In: Bove AA, Lowenthel DT (eds) *Exercise Medicine: Physiological Principles and Clinical Applications.* Academic Press, New York

Shaffstall RM, Burton RR (1979). Evaluation of assisted positive-pressure breathing on $+G_z$ tolerance. *Aviat Space Environ Med* 50: 820–824

Shaffstall RM, Burton RR (1980). Evaluation of a uniform pressure anti-G suit concept. *Aerosp Med Assoc Preprints,* pp 96–97

Simonson E, Lind AR (1971). Fatigue in static work. In: Simonson E (ed) *Physiology of Work Capacity and Fatigue.* Charles C Thomas, Springfield, IL

Spence DW, Parnell MJ, Burton RR (1981). Abdominal muscle conditioning as a means of increasing tolerance to $+G_z$ stress. *Aerosp Med Assoc Preprints,* pp 148–149

Stull GA, Clarke DH (1971). Patterns of recovery following isometric and isotonic strength decrement. *Med Sci Sports* 3: 135–139

Tamir A, Burton RR, Forster EM (1988) Optimum sampling times for maximum blood lactate levels after exposures to sustained $+G_z$. Aviat Space Environ Med 59: 54–56

Tesch PA, Balldin UI (1984). Muscle fiber type composition and G-tolerance. *Aviat Space Environ Med* 55: 1000–1003

Tesch PA, Hjort H, Balldin UI (1983). Effects of strength training on G tolerance. *Aviat Space Environ Med* 54: 691–695

Wood EH, Lambert EH, Baldes EJ, Code CF (1946). Effects of accleration in relation to aviation. *Fed Proc* 5: 327–344

Wood EH, Lambert EH, Code CF (1981). Involuntary and voluntary mechanisms for preventing cerebral ischemia due to positive $(+G_z)$ acceleration. *Physiol* 24: S-33–36

11

Effects of Weightlessness on Human Fluid and Electrolyte Physiology

Carolyn S. Leach and Philip C. Johnson, Jr.

Introduction

The fluid-regulating systems of the body have been of interest to space medicine researchers since results from the earliest flights indicated significant changes in this area (Berry et al. 1966; Dietlein and Harris 1966; Lutwak et al. 1969). The virtual absence of gravity causes a decrease in posturally induced hydrostatic force in the extremities, which leads to cephalad redistribution of blood. This redistribution is thought to be responsible for most of the spaceflight-induced changes in fluid and electrolyte metabolism. Plasma volume decreases (Johnson 1979) and water and electrolyte balances become negative (Leach 1979) in space travelers. In addition to these clear-cut effects, more complex and subtle changes in renal and circulatory dynamics, endocrine function, body biochemistry, and metabolism occur during spaceflight.

Two Phases of the Adaptation Process

Studies in which weightlessness is simulated by decreasing lower-extremity hydrostatic forces (as by bed rest or water immersion) have indicated the presence of at least two phases in the adaptation of the fluid and electrolyte homeostatic systems to microgravity (Leach et al. 1983). The "acute" phase is believed to occur within a few hours of attaining weightlessness. Since it has been difficult for astronauts to perform experiments early in a flight, most of the evidence for existence of this phase has come from simulation studies. Bed-rest studies (Leach et al. 1983; Nixon et al. 1979) have shown that central venous pressure (CVP) increases as early as 5 min after bed rest begins (Nixon et al. 1979). This is followed by an increase in the size of the left ventricle, but there is no change in cardiac output or arterial pressures (Nixon et al. 1979). The increased CVP is thought to be interpreted physiologically as an increase in total blood volume. Glomerular filtration rate (GFR) decreases by about 2 hr and effective renal plasma flow (ERPF) by 4 hr, but both return to pre-bed-rest levels by 8 hr. Plasma aldosterone and antidiuretic hormone (ADH) decrease between 1 and 6 hr after the beginning of bed rest (Leach et al. 1983; Nixon et al. 1979).

The transient acute phase, found in simulation studies and confirmed by recent Spacelab data to be discussed below, leads to a later "adaptive" phase. Evidence for the existence of the adaptive phase has come from blood and urine samples taken in-flight during Gemini, Apollo, Skylab, and Spacelab missions.

Early Spaceflight Findings

Data from limited in-flight samples, along with preflight and postflight measurements of many physiological parameters, provided evidence that mass is lost, water balance becomes negative, electrolytes and certain minerals are depleted, and cardiovascular deconditioning occurs as a result of weightlessness (Berry et al. 1966; Hoffler 1977; Leach et al. 1975). Fluid, potassium, and nitrogenous compounds were apparently lost from cells as well as from blood (Leach et al. 1975). Levels of some of the hormones involved in regulating fluid and electrolyte balance were altered; for example, urinary ADH and aldosterone and plasma angiotensin were increased postflight. Plasma volume and red cell mass decreased, and orthostatic tolerance and exercise capacity were reduced (Hoffler and Johnson 1975).

Skylab

Experiments for Skylab were planned to document the time course of known physiological changes and to measure additional parameters during long flights. Intake of fluid and nutrients during flight was carefully monitored.

The first in-flight measurements of body mass were performed on Skylab (Thornton and Ord 1977). The crew members lost an average of 2.8 kg, 3.8% of preflight body mass, during flight (Leach and Rambaut 1977). About half the loss of body mass occurred during the first 2 days of flight, with the rest of the loss being more gradual but continuing throughout the missions. Depletion of water was thought to be responsible for the rapid phase of mass loss and depletion of fat and protein for the slow phase (Leach and Rambaut 1977).

Increased urinary excretion of water was expected to account for the water deficit. Surprisingly, urinary excretion decreased during the first 10 days of the missions, and free water clearance decreased slightly (Leach and Rambaut 1977). Water balance studies showed that the main cause of the net body water reduction during the first 2 days of flight was a decrease in fluid intake.

To investigate the postulated shift of fluid away from the lower extremities, the leg volume of Skylab crew members was measured by plethysmography (Thornton et al. 1977). It was estimated that in the first few days of flight, 1.8 l of fluid disappeared from the legs (Hoffler 1977), an amount considerably greater than the 600 to 800 ml redistributed by a change in body position (upright to supine or vice versa) (Sjöstrand 1953). The amount of fluid lost from the legs was almost equal to the total body fluid decrement.

The loss of so much body fluid implies that levels of other blood and tissue components are reduced also. Plasma osmolality and levels of sodium and chloride were decreased during flight, and the amounts of sodium, potassium, calcium, phosphate, and magnesium were increased in 24-hr urine pools collected in-flight. Leach and Rambaut (1977) calculated that approximately 100 meq of sodium were lost from the extracellular space. Electrolytes and other cell constituents may have been translocated from cells to blood. Plasma levels of potassium, calcium, and phosphate increased during flight.

Plasma angiotensin and urinary aldosterone and cortisol were increased over their preflight levels during the whole flight but were particularly increased at the beginning of each flight (Leach and Rambaut 1977). These hormones are released in response to stress and to changes in plasma osmolality and electrolytes. Increased angiotensin and aldosterone may have caused at least part of the increased urinary excretion of potassium, but it is unusual for high levels of aldosterone to be associated with increased sodium excretion. Urinary excretion of antidiuretic hormone (ADH) was decreased during flight, another unexpected finding because the loss of fluid would normally stimulate ADH secretion, and hyponatremia persisted in spite of the apparent reduction in ADH.

Renal function was not measured directly during the Skylab flights. Slight increases in creatinine clearance (Leach 1981), decreased urinary and plasma uric acid, and increased plasma angiotensin indicated that renal function may be affected by weightlessness.

Recent Findings from the Space Shuttle

Several experiments involving fluid and electrolyte physiology have now been performed aboard the space shuttle. Venous pressure was measured for the first time on *Spacelab 1,* 22 hr after launch (Kirsch et al. 1984). At that time venous pressure was lower than it was on the day before launch. Measurement of central and peripheral venous pressures 1 and 12 hr after landing indicated that fluid redistribution after reexposure to gravity was completed between these times. If redistribution caused by microgravity takes about the same amount of time, it is probably complete before 22 hr.

Studies of body fluid changes during spaceflight have been hampered by lack of knowledge about changes in circadian rhythm and by flight-related problems such as space adaptation syndrome or crew members being on different work/rest cycles or being unable to draw blood very soon after reaching orbit or at the same time each day.

In one experiment a mission specialist collected his urine as pools representing 5 to 26 hr, and the excretion rates of electrolytes and selected hormones were determined. The earliest change detected in this study was a transient increase in the excretion rate of ADH in the first in-flight sample. This was closely followed by a transient increase in the excretion rate of cortisol. Sodium excretion decreased on the day after the peak in cortisol excretion occurred, but later in the flight it increased. Potassium excretion increased at the same time as cortisol excretion on the first day of flight, with smaller peaks on later flight days. Some of these changes may have been caused or affected by the presence of space adaptation syndrome. On the sixth and last day of the flight, aldosterone excretion rate tripled, and cortisol and ADH excretion rates increased by lesser amounts. The excretion rates of fluid, potassium, chloride, calcium, and magnesium increased at the same time. The loss of sodium, which might be expected to result in increased aldosterone secretion, was no greater late in the flight than it had been during the preflight period.

On the Spacelab flights, blood samples were drawn 22 or more hours after launch. Aldosterone, cortisol, and ADH were measured in blood samples from Spacelab 1 by Dr. K.A. Kirsch (personal communication), and our laboratory has measured these and other hormones as well as serum osmolality, sodium, and potassium in other Spacelab experiments. The four crew members on Spacelab 1 who participated in experiments were on two different work/rest cycles, but their blood samples were taken at the same clock time, and studies of the circadian rhythms of several urinary variables showed that the circadian rhythm of metabolic functions did not change (Leach, Johnson, & Cintrón 1985). Blood samples were obtained from four mission and payload specialists on Spacelab 2 and two mission specialists on Spacelab 3. The crew members on Spacelab 2 were on two different work/rest cycles, and during flight they collected blood samples during the postsleep activity period. This was 6:30 or 7:00 a.m. Houston time for two crew members and 5:00 or 6:00 p.m. for the other two. Because of differences in sample collection times, one must be cautious in interpreting the results, but the small number of subjects and time points in any one experiment makes it desirable to examine the results of these three experiments together.

The combined results for all three Spacelab studies (Leach et al. 1985) showed that hyponatremia developed within 20 hr after the onset of weightlessness and continued throughout the flights, and hypokalemia developed by 40 hr. Serum potassium returned to preflight levels later and then increased. Serum chloride was decreased on most in-flight days on which it was measured, but it immediately returned to preflight levels on landing day.

Antidiuretic hormone, which increased transiently in urine in the shuttle experiment described above and decreased in urine during Skylab flights (Leach and Rambaut 1977) and in plasma during bed rest (Nixon et al. 1979), was increased in plasma throughout the Spacelab flights. Aldosterone decreased by 40 hr, but after 7 days it had reached preflight levels. Angiotensin I was elevated after 2 days in flight. Cortisol increased early but decreased later in the flight. Adrenocorticotrophic hormone was increased until the seventh day, when levels of cortisol and aldosterone returned to or surpassed baseline.

Current Problems

The changes that occur in human fluid and electrolyte physiology during the acute and adaptive phases of adaptation to spaceflight are summarized in Tables 11.1 and 11.2. A number of questions remain to be answered.

At a time when plasma volume and extracellular fluid volume are contracted and salt and water intake is unrestricted, ADH does not correct the volume deficit and serum sodium decreases. Change in secretion or activity of a natriuretic factor during spaceflight is one possible explanation.

Recent identification of a polypeptide hormone produced in cardiac muscle cells which is natriuretic, is hypotensive, and has an inhibitory effect on renin and aldosterone secretion (Atarashi et al. 1984; Palluk et al. 1985) has renewed interest in the role of a natriuretic factor. The role of this atrial natriuretic factor (ANF) in both long- and short-term variation in extracellular volumes and in the inability of the kidney to bring about an escape from the sodium-retaining state accompanying chronic cardiac dysfunction makes it reasonable to look for a role of ANF in the regulation of sodium during exposure to microgravity. Prostaglandin E is another hormone that may antagonize the action of ADH (Anderson et al. 1976). Assays of these hormones will be performed on samples from crew members in the future.

TABLE 11.1. Acute phase of actual or simulated microgravity effects on fluid and electrolyte physiology

Cardiovascular effects (bed rest)
  Increased central venous pressure
  Increased size of left ventricle
Renal effects (bed rest)
  Decreased GFR
  Decreased ERPF
Endocrine system changes (Spacelab)
  Increased plasma cortisol
  Decreased plasma angiotensin I

TABLE 11.2. Adaptive phase of
microgravity effects on fluid and electrolyte
physiology, compared with preflight.

Mass loss
  Water
  Protein
  Fat
Changes in fluid volumes in major body
  compartments
  Decrease in lower body
  Increase in upper body
  Decrease in intracellular water
  Decrease in extracellular fluid
Negative fluid balance
  Decreased fluid intake
  Increased evaporative water loss
  Decreased renal excretion of water
  Slightly decreased free water clearance
  Decreased total body water
Electrolyte balance
  Decreased exchangeable body potassium
  Decreased sodium in extracellular space
  Increased excretion of sodium
  Increased excretion of potassium
Blood levels of electrolytes
  Increased potassium
  Decreased sodium
  Decreased chloride
  Decreased osmolality
Endocrine system changes
  Increased plasma angiotensin I
  Increased urinary aldosterone
  Increased urinary cortisol
  Increased plasma ADH
  Decreased urinary ADH
Renal function
  Decreased plasma and urinary uric acid
  Increased creatinine clearance

Cardiovascular intolerance to standing, found to occur immediately after landing in many astronauts, is thought to be related to loss of fluid and electrolytes during weightlessness. In the space shuttle, reentry acceleration is experienced in the head-to-foot direction because the crew members are sitting upright. The gravitational force in that direction is usually about 1.2 times the normal 1.0 G (Nicogossian and Parker 1982). The rapid increase in accleration forces during reentry would be expected to pull body fluids toward the legs. If there has been substantial loss of body fluid, fluid volume in the upper part of the body may decrease enough to cause cardiovascular symptoms. Some of these symptoms might be alleviated if fluid and electrolyte metabolism were fully re-adapted to earth's gravity before landing. Attempts have been made, with some success, by investigators in the United States and Soviet space programs to prevent or ameliorate orthostatic intolerance.

One method that has been used to counter the orthostatic intolerance is fluid and electrolyte loading. If fluid and electrolytes are replaced, blood volume should begin to increase and blood pressure should approach preflight levels. This should be done before exposure to the increased acceleration during the deorbit period. It is now standard practice for U.S. astronauts to consume the equivalent of a liter of physiological saline solution in the form of water and salt tablets before landing is initiated. This practice has been shown to be effective in reducing the severity of symptoms of cardiovascular deconditioning (Bungo et al. 1985). Similar countermeasures have been used by cosmonauts on Soyuz missions (Grigoriev 1983).

Another approach to prevention of postflight orthostatic intolerance is the use of lower-body negative pressure (LBNP) during flight to bring more fluid into the legs; this has been used with some success in the Soviet space program (Grigoriev 1983).

There is now considerable indirect evidence that renal function is altered during weightlessness (Leach, Johnson, & Cintrón 1985), but direct measurements of renal function have been done only in bed-rest studies. Renal function tests will be performed in conjunction with measurement of hormones, electrolytes, plasma volume, and other factors on Spacelab missions in the future. Intake of food and water will be measured throughout the mission, and urine will be collected void by void. Blood samples will be taken at intervals, beginning at 3 hr after launch, and the first renal function test will start at 3.5 hr. A catheter to measure central venous pressure will be inserted before launch and removed 12 hr into the flight. Plasma volume and extracellular fluid will be measured on the second and sixth days of flight. These integrated experiments are expected to provide information important for understanding what happens in both phases of the fluid and electrolyte response to weightlessness.

Acknowledgements. The authors thank Dr. Jane Krauhs and Mrs. Sharon Jackson for assistance with preparation of the manuscript.

REFERENCES

Anderson RJ, Berl T, McDonald KM, Schrier RW (1976). Prostaglandins: Effects on blood pressure, renal blood flow, sodium and water excretion. *Kidney Int* 10: 205–215

Atarashi K, Mulrow PJ, Franco-Saenz R, Snajdar R, Rapp J (1984). Inhibition of aldosterone production by an atrial extract. *Science* 224: 992–994

Berry CA, Coons DO, Catterson AD, Kelly GF (1966). Man's response to long-duration flight in the Gemini spacecraft. In: *Gemini Midprogram Conference,* February 23–25, 1966. Johnson Space Center, Houston, TX, pp 235–261

Bungo MW, Charles JB, Johnson PC Jr. (1985). Cardiovascular deconditioning during space flight and the use of saline as a countermeasure to orthostatic intolerance. *Aviat Space Environ Med* 56: 985–990

Dietlein LF, Harris E (1966). Experiment M-5, bioassays of body fluids. In: *Gemini Midprogram Conference,* February 23–25, 1966. Johnson Space Center, Houston, TX, pp 403–406

Grigoriev AI (1983). Correction of changes in fluid-electrolyte metabolism in manned space flights. *Aviat Space Environ Med* 54: 318–323

Hoffler GW (1977). Cardiovascular studies of U.S. space crews: An overview and perspective. In: Hwang NHC, Normann NA (eds) *Cardiovascular Flow Dynamics and Measurements.* University Park Press, Baltimore, MD, pp 335–363

Hoffler GW, Johnson RL (1975). Apollo flight crew cardiovascular evaluations. In: Johnston RS, Dietlein LF, Berry CA (eds) *Biomedical Results of Apollo,* NASA SP-368. National Aeronautics and Space Administration, Washington, DC, pp 227–264

Johnson PC (1979). Fluid volumes changes induced by spaceflight. *Acta Astronaut* 6: 1335–1341

Kirsch KA, Röcker L, Gauer OH, Krause R, Leach C, Wicke HJ, Landry R (1984). Venous pressure in man during weightlessness. *Science* 225: 218–219

Leach CS (1979). A review of the consequences of fluid and electrolyte shifts in weightlessness. *Acta Astronaut* 6: 1123–1135

Leach CS (1981). An overview of the endocrine and metabolic changes in manned space flight. *Acta Astronaut* 8: 977–986

Leach CS, Alexander WC, Johnson PC (1975). Endocrine, electrolyte, and fluid volume changes associated with Apollo missions. In: Johnston RS, Dietlein LF, Berry CA (eds) *Biomedical Results of Apollo,* NASA SP-368. National Aeronautics and Space Administration, Washington, DC, pp 163–184

Leach CS, Chen JP, Crosby W, Johnson PC, Lange RD, Larkin E, Tavassoli M (1985). *Spacelab 1 Hematology Experiment (INS103): Influence of Space Flight on Erythrokinetics in Man,* NASA TM 58268. Johnson Space Center, Houston, TX

Leach CS, Johnson PC, Cintron NM (1986). The regulation of fluid and electrolyte metabolism in weightlessness. In: Hunt J (ed) *Proceedings of the 2nd International Conference on Space Physiology,* Toulouse, France, November 20–22, 1985, ESA SP-237. European Space Agency, Paris, France, pp 31–36

Leach CS, Johnson PC, Suki WN (1983). Current concepts of space flight induced changes in hormonal control of fluid and electrolyte metabolism. *Physiologist* 26: S-24–S-27

Leach CS, Rambaut PC (1977). Biochemical responses of the Skylab crewmen: An overview. In: Johnston RS, Dietlein LF (eds) *Biomedical Results from Skylab,* NASA SP-377. National Aeronautics and Space Administration, Washington, DC, pp 204–216

Lutwak L, Whedon GD, Lachance PH, Reid JM, Lipscomb HS (1969). Mineral, electrolyte and nitrogen balance studies of the Gemini VII fourteen-day orbital space flight. *J Clin Endocrinol* 29: 1140–1156

Nicogossian AE, Parker JF Jr (1982). *Space Physiology and Medicine,* NASA SP-447. National Aeronautics and Space Administration, Washington, DC, p 40

Nixon JV, Murray RG, Bryant C, Johnson, RL Jr, Mitchell JH, Holland OB, Gomez-Sanchez C, Vergne-Marini P, Blomqvist CG (1979). Early cardiovascular adaptation to simulated zero gravity. *J Appl Physiol: Respirat Environ Exercise Physiol* 46: 541–548

Palluk R, Gaida W, Hoefke W (1985). Atrial natriuretic factor. *Life Sci* 36: 1415–1425

Sjöstrand T (1953). Volume and distribution of blood and their significance in regulating the circulation. *Physiol Rev* 33: 202–228

Thornton WE, Hoffler GW, Rummel JA (1977). Anthropometric changes and fluid shifts. In: Johnston RS, Dietlein LF (eds) *Biomedical Results from Skylab,* NASA SP-377. National Aeronautics and Space Administration, Washington, DC, pp 330–338

Thornton WE, Ord J (1977). Physiological mass measurements in Skylab. In: Johnston RS, Dietlein LF (eds) *Biomedical Results from Skylab,* NASA SP-377. National Aeronautics and Space Administration, Washington, DC, pp 175–182

Part IV Comparative Physiology

12

Aquatic Life and the Transition to Terrestrial Life

PIERRE DEJOURS

The question of how life on earth originated is a delicate one. If the formation of simple organic molecules seems to be explainable, since several organic compounds my be synthetized in vitro from simple components, the increased degree of complexity of these molecules and their mode of organization into living units capable of self-reproduction is very much debated. It is generally accepted that the aquatic milieu was the first abode of metazoans. I will assume in this presentation that the first metazoans were aquatic and that their descendants colonized the land.

The important consequence of this view is that terrestrial animals have to be compared with aquatic animals and not the other way around, an attitude which is not yet common. The approach which consists of comparing the characteristics, mainly physiological, of aquatic animals with those of terrestrial animals comes from the fact that most of the physiological functions were first extensively studied in terrestrial animals and were thus taken as references. But the real reference animals are aquatic. Of course, we know examples of terrestrial animals which have reinvaded the water, e.g., the ancestors of the cetaceans and of the aquatic reptiles; in these particular cases, the terrestrial forebears are the references. Some amphibians may very well show a recapitulation of the evolutionary processes. With metamorphosis, amphibians generally become terrestrial, some being able to live relatively independently of bodies of water, but in a humid air. Some commonly return to their aquatic mode of life, but others do so less often except for one key function—the laying of their eggs, which typically develop in water.

Water and air are, in all respects, very different as life-supporting media (Table 12.1). It must be noted that the media are an unbuffered fresh water and open air. In fact, some waters, not only seawater, can differ markedly from unbuffered fresh water; some gas pockets, such as underground gas collections, may differ from the open atmosphere. The table is, moreover, incomplete in that it does not show the properties of water and air regarding mechanical vibration and electromagnetic waves,

TABLE 12.1.

	Water	Air	Air / Water	Consequences
O_2 capacitance at 18°C	+'	++	29	O_2 and CO_2 partial pressures
CO_2 capacitance at 18°C	++	++	1	Acid-base balance
Mechanical properties at 18°C				Work of breathing
Viscosity	+++	+	1/60	Skeleton
Density	++++	+	1/800	Circulation
Properties for heat transfer at 15°C				
Heat capacity	++++	+	1/3400	Body temperature
Heat conductivity	++	+	1/24	
NH_3 capacitance at 20°C	+++	++	1/750	Amino acid, purine, and pyrimidine catabolism

characteristics very important for audition, vision, and electric sensitivity. Here I will mainly concentrate on respiration and acid-base balance.

Fig. 12.1 (left) shows the concentrations of O_2 and CO_2 as functions of their partial pressures in fresh unbuffered water. The figure illustrates a very important property: the coefficient of solubility, the slope of the line of concentration vs. partial pressure, which is much lower for O_2 than for CO_2. It follows (Rahn 1966) that for a given CO_2 loading and O_2 unloading in water, the corresponding partial pressure change of CO_2 is much lower than that of O_2, and this is quite apparent on a PCO_2 vs. PO_2 diagram (Fig. 12.2). On this diagram, I and E designate the partial pressures of the two gases in inspired water (here a water equilibrated with air near sea level) and expired water. The higher the O_2 extraction coefficient, the lower the PE_{O_2}. If the O_2 extraction coefficient were 100%, PE_{O_2} would be null; and in this water at 15°C, the expired PCO_2 would be 5.5 Torr, an unrealistic absolute limit. Generally, real waters, such as seawater, are buffered (Fig. 12.3); when these waters are loaded with CO_2, the change in PCO_2 due to CO_2 loading is moderated because some CO_2 is fixed as bicarbonate. As a consequence (Fig. 12.4), in the case of seawater, the PCO_2 vs. PO_2 line is as in 2, instead of 1, which is for distilled water.

The ratio $dCCO_2/dPCO_2$ has the dimension of a coefficient of solubility, as used in Henry's law. It is an effective solubility which is nowadays generally called capacitance, and its exact value for a given water requires a precise knowledge of the temperature, the salinity, the composition, and the pH of the milieu. Note that the term "capacitance" is also used for the "solubility" of a gas species in a gas phase, and for the relation between O_2 and CO_2 concentrations as functions of their tensions in body fluids. It may be added that some waters contain much more buffer than seawater and the fresh water of Fig. 12.3 (see Dejours 1981). This fact explains why

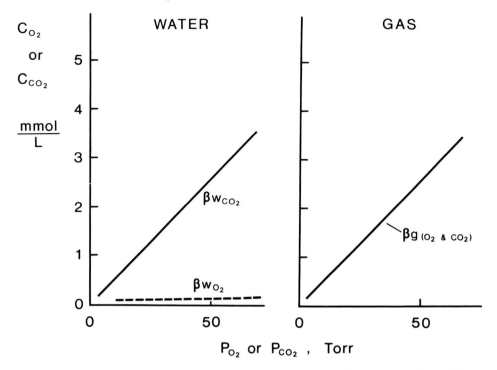

FIGURE 12.1. Concentrations of O_2 and CO_2, Co_2 and Cco_2, vs. their partial pressure, Po_2 and Pco_2, in distilled or unbuffered water and in air at 15°C. The slopes correspond to the O_2 and CO_2 capacitance coefficients of these milieus. (From Dejours 1981.)

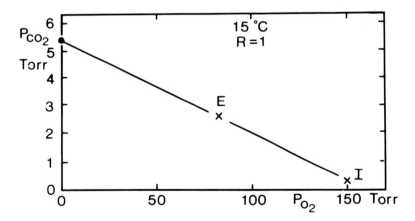

FIGURE 12.2. Pco_2 vs. Po_2 diagram in distilled or unbuffered fresh water. I designates the inspired water equilibrated with air near sea level, and E designates the expired water. The higher the O_2 extraction coefficient, the further to the left the expired point (see text). R is the respiratory quotient.

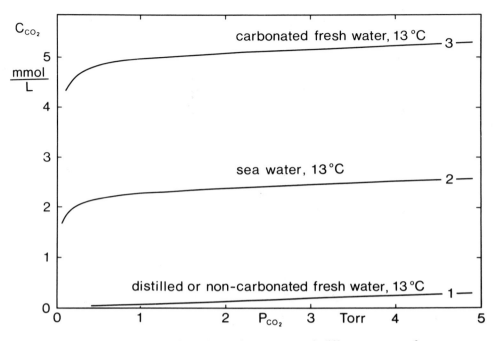

FIGURE 12.3. Carbon dioxide absorption curves of different types of waters. (From Dejours et al. 1968.) See text.

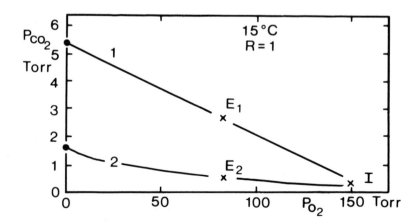

FIGURE 12.4. P_{CO_2} vs. P_{O_2} diagram in distilled or unbuffered fresh water (1) and in seawater (2). I, inspired water; E, expired water. The increase of P_{CO_2} accompanying the fall in P_{O_2} in the external respiration process is much smaller in seawater (point E_2) than in unbuffered water (point E_1). (From Dejours 1978.)

blood P_{CO_2} may be much lower in breathers of buffered waters than in breathers of unbuffered waters.

Air breathers live in a milieu which contains 20 to 40 times more oxygen than water for the same partial pressure, whereas the CO_2 capacitance coefficient in air is similar to its value in distilled water (Fig. 12.1, right). It follows that if the ventilation were the same in air breathers and water breathers, their expired P_{CO_2} would be similar. But since air breathers live in a O_2-rich milieu compared with water, they can breathe less and actually do (Rahn 1966), and as a result expired P_{CO_2} is increased. On the P_{CO_2} vs P_{O_2} diagram of Fig. 12.5 (Rahn and Fenn 1955), the lines I-E correspond to inhaled and exhaled milieus, air for the human and water for the dogfish; and the arterial blood P_{CO_2} reflects the expired P_{CO_2} (even if it may differ somewhat because of the organization of the external respiration exchangers).

Fig. 12.5 suggests that it would be quite easy to differentiate between an air breather and a water breather. Actually it is not always so obvious because the blood P_{CO_2} of cold-blooded air breathers, for instance, reptiles, at the same temperature as the water breathers, has much lower

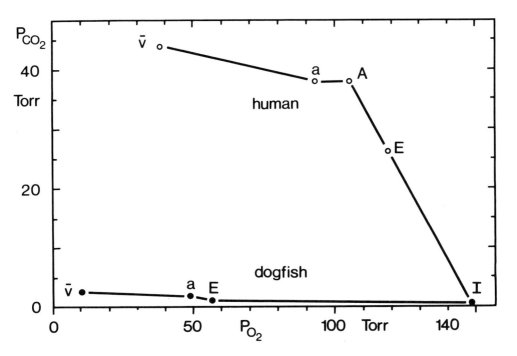

FIGURE 12.5. P_{CO_2} vs. P_{O_2} diagram. Symbols: 1, inspired milieu; E, expired milieu; A, alveolar gas; a, arterial blood; $\bar{v}$, mixed venous blood. The data for the dogfish are taken from Baumgarten-Schumann and Piiper (1968); those for the human are from Farhi and Rahn (1960).

P_{CO_2}'s than the human. Indeed,· in the air-breathing turtle *Pseudemys scripta elegans* (Jackson et al. 1974), P_{CO_2} falls markedly with the fall of temperature because of a relative hyperventilation (increase of the ratio $\dot{V}_E/\dot{M}_{CO_2}$). In this turtle, Pa_{CO_2} is only 14 Torr at 10°C and 32 Torr at 30°C, a change which is related to the problem of acid-base balance. Taking into account the effect of temperature on body P_{CO_2}, and also the fact that unmixed arterial blood in vertebrates exists only in mammals, birds, and fish, the difference between air breathers and water breathers remains clear if attenuated, namely, that the carbon dioxide partial pressure is higher in the air breathers than in water breathers.

The pH values may differ considerably among species, but there is no systematic difference between air and water breathers. Since air breathers have higher P_{CO_2}'s than water breathers and since the two groups have similar pH's, it follows that [HCO_3^-] is higher in air breathers than in water breathers. This is clearly shown by Fig. 12.6. If we accept that air breathers evolved from water breathers, then we must say that air

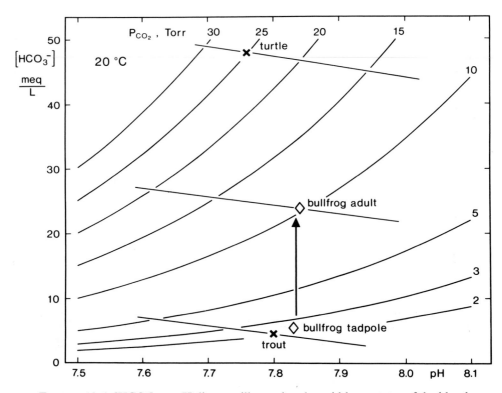

FIGURE 12.6. [HCO_3^-] vs. pH diagram illustrating the acid-base status of the blood of the turtle with reference to the trout (x) and the adult with reference to the tadpole bullfrog (◇). (From Howell et al. 1970.)

breathers are in a hypercapnic status, with reference to water breathers. Then, since their pH's are alike, it follows that the air breathers are in a state of compensated hypercapnic acidosis. This view is reinforced by what happens in the metamorphosing amphibian, as shown in Fig. 12.6 for the larval and adult bullfrog. Many examples of the change in acid-base balance brought about by the transition from aquatic to aerial breathing may be found in Rahn and Garey (1973) and in Rahn and Howell (1976).

The increase of $[HCO_3^-]$ with the change from water to air breathing raised other problems which had to be solved for a successful terrestrial invasion, in particular ionoregulation, itself inseparable from the necessity for water conservation and the excretion of solutes. It is not possible to discuss here these questions which have received considerable experimental attention. The only fact I will mention is that most aquatic animals are mainly ammoniotelic; that is, the N atoms of the amino acids and the purine and pyrimidine bases are excreted as ammonia, NH_3, or more exactly as NH_3/NH_4^+. This is made possible by the fact that NH_3 is extremely soluble in water. The NH_3 capacitance in air is comparatively very low, and in terrestrial animals, organic nitrogen is mainly excreted as uric acid which may precipitate, or as urea which is highly water-soluble. However, elasmobranchs form some urea which reaches high concentrations in their blood. There it plays an important role in keeping the osmotic pressure of these animals at a value close to seawater osmolarity. But in teleosts, which are hypo-osmotic to seawater, the blood-urea concentration is similar to that of mammals.

Generally the relative mass of the skeleton of aquatic dwellers, who benefit from an apparent null gravity, is constant whatever the body size. In terrestrial animals, the skeletal percentage—cuticle of spiders, shell of avian eggs, bones of mammals and birds—is a size-dependent. The heavier a terrestrial animal, the higher its relative skeletal mass—for example, 4% in a shrew, 25% in an elephant. However, in some aquatic animals, like the freshwater mollusks and the whales, the skeleton development is also size-dependent (Anderson et al. 1979).

Gravity is a stress for blood circulation. The hydrostatic pressure of the arterial blood which perfuses the brain is the same whatever the body mass and the size of the animal. As a consequence, the blood pressure in the thoracic aorta, and as well in the distal arteries of the legs of standing animals, is much higher in tall animals than in small ones.

Table 12.1 shows that heat dissipation is much less easy in air than in water. Actually the temperature of aquatic animals is close to that of the ambient water, with two exceptions: (1) In some swimming muscles of the tuna, the temperature is higher than in the rest of their body, and to a certain extent this localized hyperthermia is independent of the thermal change of the environment. (2) Whales and seals have a core temperature typical of hemeothermic vertebrates, but the viscera are insulated from the water by thick blubber. All other homeotherms are terrestrial animals, and there is little doubt that their genesis was made possible by the low-heat-dissipative properties of air.

Many of the examples given here are taken from among aquatic and terrestrial vertebrates. However, the points made concern most aquatic animals, on the one hand, and most terrestrial animals, on the other. For example, it is very clear that all air breathers have a higher body fluid P_{CO_2} than water breathers. Most of the completely aquatic animals are ammoniotelic, whereas terrestrial animals are not. Problems of water conservation concern all terrestrial animals. Whether one compares reptiles, birds, and mammals to fish; frog to tadpole; insects and arachnids to aquatic arthropods; terrestrial crustacea to aquatic crustacea; terrestrial mollusks to aquatic mollusks, these considerations are generally valid. Many physiological traits are convergent, are related to the environmental characteristics, and transcend the zoological divisions of the animal kingdom (Dejours 1979). Whether the aquatic animals are invertebrates or vertebrates, the problems raised by the invasion of land are common to all. This ecophysiological transphyletic division of the animal kingdom is very clear when one looks at completely aquatic animals and completely terrestrial animals. It may look less clear for intermediary animals, either because they keep traits of their original abode, e.g., the whales, or because they live half in air, half in water, as some amphibians. But intermediary animals exist in all classifications of living creatures, and they do not upset the classification. On the contrary, intermediary forms may be very instructive because they may shed light upon the evolutionary process by which some aquatic animals were able to invade land.

Summary

The physicochemical properties of water and air differ by the capacitance coefficients of O_2. CO_2, and NH_3 in the two media; by the specific mass and viscosity; by the transmission of mechanical and electromagnetic waves; by the heat capacity and conductivity; by the water evaporation in air; and by the latent heat accompanying the change of water phase.

To obtain a given amount of oxygen, water breathers exhale a water whose P_{O_2} is some tens of Torrs lower than that of inhaled water. To this inspired-to-expired P_{O_2} fall, $\Delta P_{I,E_{O_2}}$, corresponds an expired-to-inspired P_{CO_2} increase, $\Delta P_{E,I_{CO_2}}$, of 1 Torr or less, since the water-CO_2 capacitance coefficient is relatively much higher than that of O_2. For air breathers, the value of $\Delta P_{E,I_{CO_2}}$ is of the same order as that of $\Delta P_{I,E_{O_2}}$; that is to say, P_{CO_2} of body fluids is much higher in air breathers than in water breathers. However, the pH's of body fluids of aquatic (low P_{CO_2}) and terrestrial (high P_{CO_2}) animals are alike; the terrestrial animals have a relatively high $[HCO_3^-]$.

Terrestrial animals have had to develop mechanisms for water conservation and concentration and excretion of salts and nitrogen wastes. Generally, wholly aquatic animals are ammoniotelic, whereas active terrestrial animals are uricotelic or ureotelic or both, ammonia being 700 times less "soluble" in air than in water. In relation to the constraints of gravity, the mass percentage of supporting tissues, skeleton or eggshell, increases with size in terrestrial animals, whereas it is generally constant in aquatic animals of differing sizes. Some principles of locomotion and of sensory mechanisms are different in terrestrial and aquatic animals. Although there are examples of localized homeothermy, for instance, brain and muscles in some fish, complete homeothermy exists only in upper vertebrates. Some of the changes necessary for a successful invasion of land are shown by the anatomical, biochemical, and physiological changes observed during the metamorphosis of an aquatic tadpole into a terrestrial, air-breathing adult amphibian.

Whether the aquatic animals are invertebrates or vertebrates, the problems raised by the invasion of land are common to all. Many physiological traits are convergent, are related to the environmental characteristics, and suggest an ecophysiological transphyletic division of the animal kingdom.

REFERENCES

Anderson JF, Rahn H, Prange HD (1979). Scaling of supportive tissue mass. *Quart Rev Biol* 54: 139–148

Baumgarten-Schumann D, Piiper J (1968). Gas exchange in the gills of resting unanesthetized dogfish *Scyliorhinus stellaris*. *Respir Physiol* 5: 317–325

Dejours P (1978). Carbon dioxide in water- and air-breathers. *Respir Physiol* 33: 121–128

Dejours P (1979). La vie dans l'eau et dans l'air. *Pour la Science* 20: 87–95

Dejours P (1981). *Principles of Comparative Respiratory Physiology,* 2nd ed. Elsevier/North-Holland, Amsterdam, 265 pp

Dejours P, Armand J, Verriest G (1968). Carbon dioxide dissociation curves of water and gas exchange of water breathers. *Respir Physiol* 5: 23–33

Farhi LE, Rahn H (1960). Dynamics of change in carbon dioxide stores. *Anesthesiology* 21: 604–614

Howell BJ, Baumgardner FW, Bondi K, Rahn H (1970). Acid-base balance in cold-blooded vertebrates as a function of body temperature. *Am J Physiol* 218: 600–606

Jackson DC, Palmer SE, Meadow WL (1974). The effects of temperature and carbon dioxide breathing on ventilation and acid-base status of turtles. *Respir Physiol* 20: 131–146

Rahn H (1966). Aquatic gas exchange: Theory. *Respir Physiol* 1: 1–12

Rahn H, Fenn WO (1955). *A Graphical Analysis of the Respiratory Gas Exchange. The O_2-CO_2 Diagram.* The American Physiological Society, Washington, DC, 38 pp

Rahn H, Garey WF (1973). Arterial CO_2, O_2, pH and HCO_3^- values of ectotherms living in the Amazon. *Am J Physiol* 225: 735–738

Rahn H, Howell BJ (1976). Bimodal gas exchange. In: Hughes GM (ed) *Respiration of Amphibious Vertebrates*. Academic Press, London, pp 271–285

13

Gas-Exchange Efficiency of Fish Gills and Bird Lungs

Johannes Piiper

Model for Efficiency of Gas-Exchange Organs

A simplified general functional schema of gas-exchange organs (Piiper 1982; Piiper and Scheid 1975, 1982a) is shown in Fig. 13.1. Medium, with flow date $\dot{V}$, and blood, with flow rate $\dot{Q}$, come into gas-exchange contact, leading to O_2 uptake by blood and CO_2 output from blood. The better the gas-transfer conditions, the closer come Po_2 and Pco_2 in medium and blood. A complete "overlap" (arterial equal inspired, venous equal expired partial pressure), as shown in case 4 of Fig. 13.1, can only be reached with an ideal countercurrent system, i.e., lacking diffusion limitation, and with perfect matching of blood and gas conductances (Piiper and Scheid 1982b, 1984).

It is well known that in mammalian lungs (Fig. 13.2), usually no complete blood-gas equilibration is reached: an alveolar-to-arterial Po_2 difference and an arterial-to-alveolar Pco_2 difference (corresponding to case 1 in Fig. 13.1) are present. These gas-to-blood differences are customarily attributed to ventilation-perfusion inequality, to shunt, and to diffusion limitation. In the absence of these limiting or efficiency-reducing factors, an equality of Pco_2 and Po_2 in blood and gas leaving the lung is achieved (case 2 of Fig. 1). An "overlap" like the ones in cases 3 and 4 of Fig. 1 does not occur in mammalian lungs in steady-state gas exchange.

The same appears to be true for pulmonary gas exchange in reptilian and amphibian lungs, but experimental data are scarce. Only in fish gills and in avian lungs might a substantial "overlap" of gas and blood Po_2 and/or Pco_2, corresponding to case 3 in Fig. 13.1, occur, but frequently the behavior is that of case 2 or 1; i.e., the gas-exchange efficiency is reduced to levels encountered in mammals.

Fish Gills

Po_2 in arterial blood higher than in expired water has been repeatedly reported, particularly in elasmobranchs (Baumgarten-Schumann and Piiper 1968; Piiper and Schumann 1967). This finding has been attributed

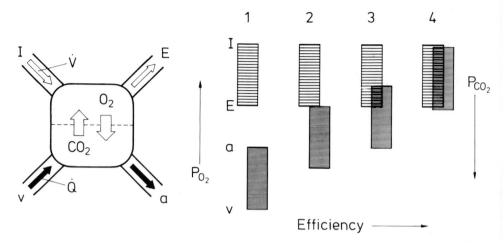

FIGURE 13.1. Efficiency of gas-exchange organs. *Left:* Model with ventilation ($\dot{V}$) and blood flow ($\dot{Q}$). *Right:* Columns delimiting the range of P_{O_2} and P_{CO_2} in air (striped) and in blood (shaded). I, E, a, and v denote inspired, expired, arterial, and (mixed) venous. Cases 1 to 4 exemplify increasing gas-exchange efficiency.

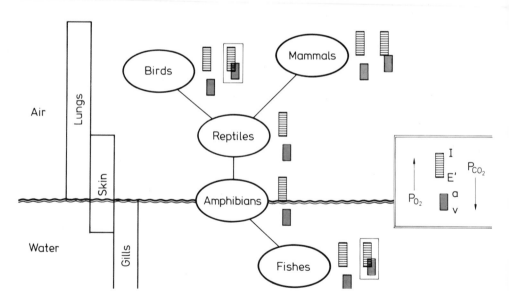

FIGURE 13.2. Medium/blood P_{O_2} and P_{CO_2} relationships. Schematized pedigree of vertebrates, their respiratory organs, and respiratory media. Meaning of columns, see Fig. 1. E′, end-expired. It should be noted that for application to skin breathing, multimodal breathing, or periodic breathing, the schema for P_{O_2} and P_{CO_2} needs modifications.

to the functional gill structure allowing countercurrent gas exchange between the water current flowing between the secondary gill lamellae and the blood flow in the secondary lamellae, as shown schematically in Fig. 13.3 (cf Piiper and Scheid 1982a, 1982b, 1984). However, in another elasmobranch, *Squalus suckleyi,* arterial Po_2 higher than expired Po_2 has not been found (Lenfant and Johansen 1966); neither has this been observed in the well-studied teleost *Salmo gairdneri* (Randall et al. 1967). Apparently, arterial Po_2 higher than expired Po_2 in fish may occur, but is not a constant feature.

In a recent study (Piiper et al. 1986), it was attempted to quantitatively explain the gas-exchange efficiency of fish gills on the basis of the countercurrent model, combining physiological measurements, physio-chemical data, and morphometrical dimensions of gills (Hughes et al. 1986). In essence, an effective diffusing capacity for O_2 (D_{eff}) was calculated from physiological measurements and was compared both with the O_2 diffusing capacity calculated from morphometrical dimensions and with solubility and diffusivity data from the literature (D_{morph}).

The effective physiological O_2 diffusing capacity (D_{eff}) was calculated from O_2 uptake, $\dot{M}o_2$, and the mean water-to-blood Po_2 difference, $(P_w-P_b)_{O_2}$, obtained by adjusting the Bohr integration technique to the countercurrent system (Piiper et al. 1977):

$$D_{eff} = \dot{M}o_2/(P_w-P_b)_{O_2} \qquad (13.1)$$

The resistance to diffusive O_2 uptake ($= 1/D_{morph}$) in the fish gill model was considered as the sum of two resistances in series, the resistance

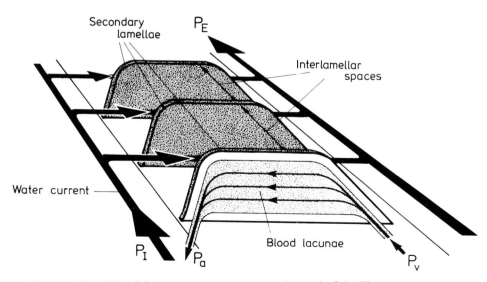

FIGURE 13.3. Model for countercurrent gas exchange in fish gills.

offered by the blood/water tissue barrier ($=1/D_m$) and the resistance residing in the water contained in the interlamellar space ($= 1/D_w$):

$$1/D_{morph} = 1/D_m + 1/D_w \qquad (13.2)$$

D_m was obtained from the surface area of secondary lamellae (F), the harmonic mean water-to-blood distance (s), and the diffusion constant (d) and solubility (α) of O_2 in tissues:

$$D_m = d \cdot \alpha \cdot F/s \qquad (13.3)$$

To estimate D_w, the interlamellar spaces were modeled as trapezoidal slits with laminar water flow and with constant Po_2 at the lateral boundaries (Scheid and Piiper 1971b). In this model, D_w is determined by (1) the number of units (= interlamellar spaces), (2) the dimensions of the units (height, width, base length, and top length (see Fig. 13.4), (3) the diffusivity and the solubility of O_2 in water, and (4) the mean water velocity (D_w increases with water flow velocity, i.e., with increasing ventilatory water flow).

The two essential simplifying assumptions of the model, constancy of Po_2 on the secondary lamella and additivity of tissue and water resistances to O_2 diffusion, have been shown to be justified for the conditions prevailing in the gills of *Scyliorhinus stellaris* and the accuracy required (Scheid et al. 1986).

For *Scyliorhinus stellaris* of 2.2- to 2.5-kg mean body mass, the following D values (in μmol/min/Torr) were estimated for the three

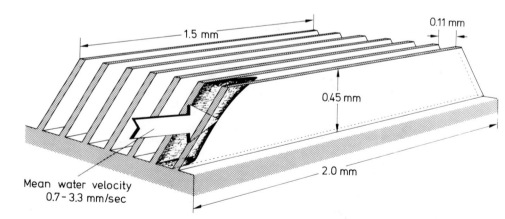

FIGURE 13.4. Morphometrical model for secondary lamellae and interlamellar spaces (after mean values determined in *Scyliorhinus stellaris* of 2.4-kg mean body mass). Seven secondary lamellae on one side of the gill filament are shown. There are about 950 filaments with about 230,000 secondary lamellae.

physiological states, resting quiescent (rq), resting alert (ra), and swimming (sw):

$$D_m: \qquad \text{rq, 3.9;} \quad \text{ra, 4.4;} \quad \text{sw, 4.4}$$
$$D_w: \qquad \text{rq, 2.1;} \quad \text{ra, 2.6;} \quad \text{sw, 3.0}$$
$$D_{morph}: \qquad \text{rq, 1.4;} \quad \text{ra, 1.7;} \quad \text{sw, 1.8} \qquad (13.4)$$
$$D_{eff}: \qquad \text{rq, 0.8;} \quad \text{ra, 1.6;} \quad \text{sw, 2.0}$$

These values show that D_m is higher than D_w, meaning that inter-lamellar water offers more resistance to diffusion of O_2 than the water-blood tissue barrier. But this relationship may vary largely from species to species as a function of the thickness of the gill epithelium and width of the interlamellar space.

Moreover, D_{eff} is lower than D_{morph} in quiescent resting fish, but in alert and swimming fish D_{eff} is close to D_{morph}. This agreement between D_{eff} and D_{morph} is rather surprising, first, because the underlying data are uncertain, and second, because the calculation of D_{eff} is based on the assumption of an idealized homogeneous model in which ventilation, blood flow, and diffusion properties of all secondary lamellae and interlamellar spaces are equal. If this were not the case, i.e., if "inho-mogeneity" were present, D_{eff} should be smaller than D_{morph}. Thus the good agreement between D_{eff} and D_{morph} may be interpreted to indicate that there is little space for gross inhomogeneity, particularly not for water or blood shunt, in the gills of *Scyliorhinus stellaris* during alertness and swimming.

The arterial P_{O_2} in the resting condition has been found to be low in many species, for example, carp (Itazawa and Takeda 1978) and lemon shark (Bushnell et al. 1982). Only in part is this due to low ventilation. It is possible that blood shunting, closure of gill areas not required for resting O_2 uptake, etc., may be involved. The low D_{eff}/D_{morph} ratio in quiescent resting *Scyliorhinus stellaris* may be due to similar mechanisms.

The mean water-to-blood P_{O_2} difference, obtained as O_2 uptake/D_{eff}, is about 50–120 Torr in *Scyliorhinus stellaris*. Since this P_{O_2} difference is largely due to diffusion, water-to-blood O_2 transfer is strongly diffusion-limited. Thus the potentially high efficiency of the countercurrent system is much diminished by diffusion limitation. But probably it is more appropriate to consider that, in view of the unfavorable properties of water as an O_2 carrier (low solubility and diffusivity for O_2, high viscosity, and high density), the presence of the potentially efficient countercurrent system is important for the economy of gas exchange in fish.

Bird Lungs

Because of high oxidative metabolism (sustained flight) and remarkable tolerance of hypoxia (Black and Tenney 1980; Lutz and Schmidt-Nielsen 1977; Shams and Scheid 1987; Tucker 1968), birds are expected to possess

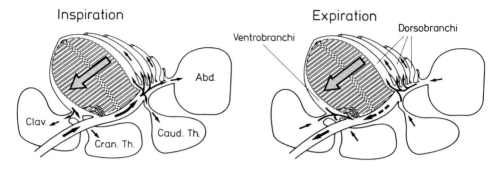

FIGURE 13.5. Schema of the avian lung–air sac system. Direction of gas flow during inspiration and expiration is indicated by arrows. Shaded arrow, airflow through the parabronchi of the lung. The clavicular (Clav.), cranial thoracic (Cran. Th.), caudal thoracic (Caud. Th.), and abdominal (Abd.) air sacs are shown.

lungs that are particularly efficient in gas exchange. The structure of the bird respiratory tract is highly specific and remarkably different from that of mammals (Duncker 1974). It consists of the air sacs, which are responsible for volume changes in the respiratory cycle, and the lungs proper, of practically constant volume (Fig. 13.5). The lungs proper are arrays of parallel parabronchi, about 1 mm wide, which mainly interconnect two sets of secondary bronchi, the dorsobronchi and the ventrobronchi. During both inspiration and expiration, air flows through the parabronchi.

AIR CAPILLARIES-BLOOD CAPILLARIES: COUNTERCURRENT-LIKE SYSTEM

From the parabronchial lumen originate air capillaries which form a network that interlaces with the blood capillary system. Anatomical evidence indicates that the prevailing flow direction in the blood capillaries is toward the parabronchial lumen (Duncker 1974). This means that the last contact of capillary blood with air in an air capillary occurs near its origin from the parabronchus (Fig. 13.6). Hence, if there exists a partial pressure gradient in the air capillary, P_{O_2} diminishing and P_{CO_2} increasing from the parabronchus to the distal end of the air capillary, the last contact of capillary blood is with gas similar to parabronchial gas (Scheid 1978). This means that in spite of the presence of diffusive gradients in the air capillaries, the overall gas transport is not affected. Thus replacement of atmospheric nitrogen by helium, leading to increased diffusivity, had no effect on gas-exchange efficiency in ducks (Burger et al. 1979).

The partial pressure profiles of Fig. 13.6 are similar to those of a countercurrent exchange system, but it must be noted that in the air

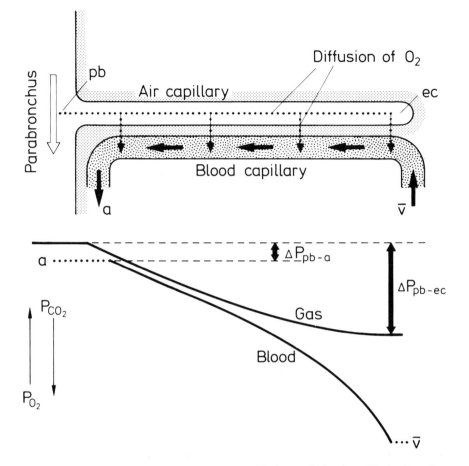

FIGURE 13.6. Gas exchange between air capillaries and blood capillaries in avian lungs. Schematic model with P_{O_2} and P_{CO_2} profiles. $\Delta P_{pb\text{-}ec}$ is the partial pressure difference between the parabronchus (pb) and the end of the air capillary (ec). $\Delta P_{pb\text{-}a}$ is the partial pressure difference between the parabronchial gas and the arterialized blood (a).

capillaries O_2 and CO_2 are transported by diffusion in a stationary gas phase, not by convection.

PARABRONCHUS-BLOOD CAPILLARIES: CROSSCURRENT SYSTEM

There is now good experimental evidence that airflow in the parabronchi occurs in the same direction during inspiration and expiration—from the dorsobronchi via parabronchi to the ventrobronchi (Bouverot and Dejours 1971; Brackenbury 1971; Bretz and Schmidt-Nielsen 1971; Scheid and Piiper 1971a) (Fig. 13.5).

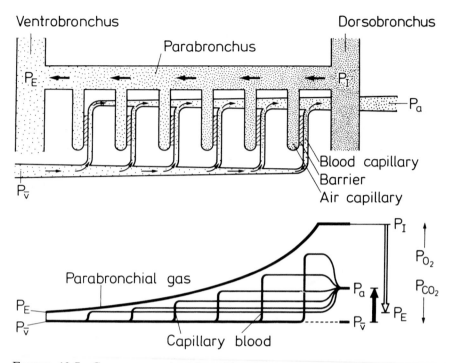

FIGURE 13.7. Crosscurrent (serial multicapillary) model for analysis of gas exchange in bird lungs. Density of shading indicates P_{O_2}. *Lower panel:* P_{CO_2} and P_{O_2} profiles.

The unidirectionality of airflow has suggested the hypothesis of a countercurrent airflow/blood-flow gas-exchange system in avian lungs (Schmidt-Nielsen 1971). However, experimental reversal of airflow through the parabronchi has been shown to leave gas-exchange efficiency unchanged, with persisting blood-gas "overlap" for CO_2, i.e., arterial P_{CO_2} lower than end-expired P_{CO_2} (Scheid and Piiper 1972). This result, together with anatomical evidence (Duncker 1974), makes it very probable that the adequate model for gas exchange in avian lungs is a serial multicapillary or crosscurrent system (Scheid 1979; Scheid and Piiper 1970) (Fig. 13.7).

Due to O_2 uptake in the air capillaries, P_{O_2} in the parabronchial gas decreases from the initial-parabronchial value, which is close to the inspired P_{O_2} (P_I), to the end-parabronchial value, which is close to the end-expired P_{O_2} (P_E). At every site along the parabronchus an equilibrium of capillary blood with air capillary gas is approached. This results in differing degrees of blood O_2 saturation in the arterialized blood according to the P_{O_2} profile in the parabronchial lumen. But the arterial blood formed by mixing may have a higher P_{O_2} (P_a) than the end-parabronchial gas (P_E) due to an admixture of highly oxygenated blood draining the initial segment of the parabronchus. In an analogous manner, the fact that P_{CO_2} is lower in arterial blood than in end-parabronchial gas is explained.

In resting birds breathing air, arterial P_{O_2} has usually been found to be considerably lower than end-expired P_{O_2} (Scheid 1979). This is explainable on the basis of "functional inhomogeneities" which are similar to those known from mammalian lungs (shunt, ventilation-perfusion inhomogeneity) and have recently been analyzed for avian lung models by Powell and Wagner (1982a, 1982b).

ENHANCED CO_2 TRANSFER: HALDANE EFFECT AND CROSSCURRENT SYSTEM

It is well known from the analysis of mammalian gas exchange that shunt and ventilation-perfusion inequalities produce much smaller blood-gas partial pressure differences for CO_2 than for O_2. This may be a reason that in birds the blood-gas "overlap" is found more regularly for CO_2 than for O_2.

But another reason could be the following remarkable mechanism which is based on the action of the Haldane effect in a crosscurrent system (Meyer et al. 1976) (Fig. 13.8). In the distal portion of the parabronchus (end-parabronchial region, ep), there is little CO_2 exchange (because parabronchial P_{CO_2} has come close to oxygenated venous blood P_{CO_2}), but O_2 uptake is only little reduced due to the shape of the O_2 dissociation curve of blood. Therefore, the effective gas-exchange ratio R is much reduced. Where R drops below a critical value (about 0.3), P_{CO_2} increases with blood-gas equilibration, because the increase of P_{CO_2} due to the Haldane effect prevails over the decrease of P_{CO_2} resulting from loss of CO_2 from blood.

Since the end-parabronchial P_{CO_2} is (approximately) equal to end-tidal P_{CO_2}, mixed venous P_{CO_2} is depressed by the Haldane effect and may reach values lower than end-expired P_{CO_2}. Indeed, this has been observed (Davies and Dutton 1975; Meyer et al. 1976). Thus, whereas in mammalian gas transport the Haldane effect is useful only insofar as it decreases venous-to-arterial P_{CO_2} differences (for equal corresponding CO_2 content differences), in bird lungs both venous and arterial P_{CO_2} are lowered in such a manner that mixed venous P_{CO_2} may decrease below end-expired P_{CO_2} (Fig. 13.9).

The reason for a particularly efficient CO_2 elimination in bird lungs is not obvious. It may provide an additional means of stabilizing the acid-base balance during stress-like exercise. But in high altitude it may contribute to aggravate the respiratory alkalosis due to hyperventilation. Indeed, in this situation a mechanism for reducing the efficiency of CO_2 exchange as compared with that of O_2 exchange may appear more useful.

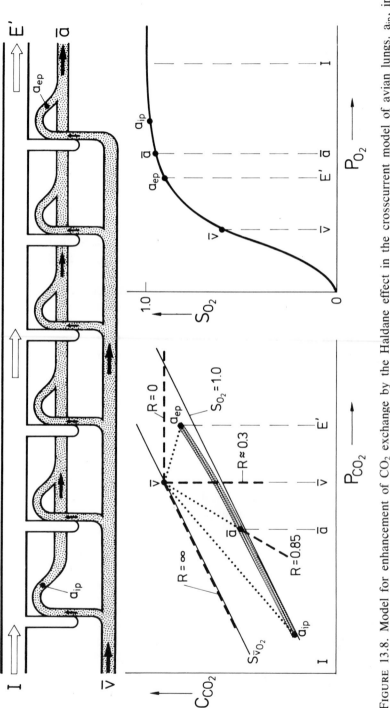

FIGURE 13.8. Model for enhancement of CO_2 exchange by the Haldane effect in the crosscurrent model of avian lungs. a_{ip}, initial-parabronchial arterial; a_{ep}, end-parabronchial arterial; $\bar{a}$, mixed arterial. *Upper panel:* Crosscurrent (serial multicapillary model). *Lower panel right:* Blood O_2 dissociation curve. *Lower panel, left:* Blood CO_2 dissociation curves for $S_{\bar{v}}O_2$ and $SO_2 = 1.0$ (straight continuous lines). Dashed lines indicate changes of P_{CO_2} and C_{CO_2} from the mixed venous value ($\bar{v}$) at certain values of the gas-exchange ratio (R). Dotted lines indicate changes of P_{CO_2} and C_{CO_2} from the mixed venous to different end-capillary or arterial values. The triple line from a_{ip} to a_{ep} indicates effective end-capillary P_{CO_2} and C_{CO_2} at different sites along the parabronchus. The sequence of the P_{O_2} and P_{CO_2} values is indicated by the symbols I, $\bar{a}$, $\bar{v}$, and E' above the abscissae.

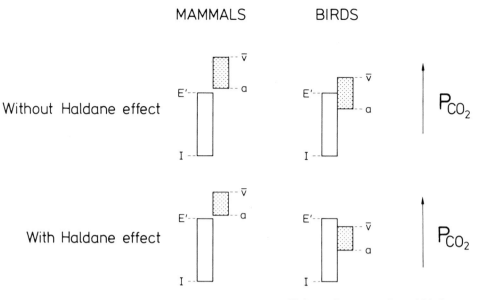

FIGURE 13.9. Haldane effect and CO_2-exchange efficiency in mammals and birds. Columns delimit the range of P_{CO_2} in air (blank) and in blood (stippled).

Summary

Efficiency of gas-exchange organs can be analyzed in terms of the relationship of P_{O_2} and P_{CO_2} in the medium (air or water) and blood entering and leaving the gas-exchange organ (i.e., in inspired and expired medium, in mixed venous and arterialized blood). A particularly enhanced efficiency is marked by arterial P_{O_2} higher than expired P_{O_2} and/or arterial P_{CO_2} lower than expired P_{CO_2}. Such relationships have been observed in fish and in birds.

The gas-exchange efficiency of the countercurrent water–blood-flow system of fish gills is considerably reduced by diffusion resistance, a great part of which resides in the water between the secondary gill lamellae. A quantitative model analysis has led to a good agreement between morphometrical data and physiological gas-exchange measurements obtained in the elasmobranch *Scyliorhinus stellaris*.

Gas exchange in bird lungs can be analyzed using the serial multicapillary or crosscurrent model in which a number of blood capillaries distributed over the length of parabronchi have gas-exchange contact with parabronchial gas. The efficacy of the system is further enhanced for CO_2 due to the Haldane effect.

In many cases the expected high efficiency of the countercurrent (fish gills) and crosscurrent system (bird lungs) is apparently not utilized.

REFERENCES

Baumgarten-Schumann D, Piiper J (1968). Gas exchange in the gills of resting unanesthetized dogfish (*Scyliorhinus stellaris*). *Respir Physiol* 5: 317–325

Black CP, Tenney SM (1980). Oxygen transport during progressive hypoxia in high-altitude and sea-level waterfowl. *Respir Physiol* 39: 217–239

Bouverot P, Dejours P (1971). Pathway of respired gas in the air sac–lung apparatus of fowl and ducks. *Respir Physiol* 13, 330–342

Brackenbury JH (1971). Airflow dynamics in the avian lung as determined by direct and indirect methods. *Respir Physiol* 13: 319–329

Bretz WL, Schmidt-Nielsen K (1971). Bird respiration: Flow patterns in the duck lung. *J Exp Biol* 54: 103–118

Burger RE, Meyer M, Graf W, Scheid P (1979). Gas exchange in the parabronchial lung of birds: Experiments in unidirectionally ventilated ducks. *Respir Physiol* 36: 19–37

Bushnell PG, Lutz PL, Steffensen JF, Oikari A, Gruber SH (1982). Increases in arterial blood oxygen during exercise in the lemon shark (*Negaprion brevirostris*). *J Comp Physiol* 147: 41–47

Davies DG, Dutton RE (1975). Gas-blood P_{CO_2} gradients during avian gas exchange. *J Appl Physiol* 39: 405–410

Duncker HR (1974). The structure of the avian respiratory tract. *Respir Physiol* 22: 1–19

Hughes GM, Perry SF, Piiper J (1986). Morphometry of the gills of the elasmobranch *Scyliorhinus stellaris* in relation to body size. *J Exp Biol* 121: 27–42

Itazawa Y, Takeda T (1978). Gas exchange in the carp gills in normoxic and hypoxic conditions. *Respir Physiol* 35: 263–269

Lenfant C, Johansen K (1966). Respiratory function in the elasmobranch *Squalus suckleyi* G. *Respir Physiol* 1: 13–29

Lutz PL, Schmidt-Nielsen K (1977). Effect of simulated altitude on blood gas transport in the pigeon. *Respir Physiol* 30: 383–388

Meyer M, Worth H, Scheid P (1976). Gas-blood CO_2 equilibration in parabronchial lungs of birds. *J Appl Physiol* 41: 302–309

Piiper J (1982). Respiratory gas exchange at lungs, gills and tissues: Mechanisms and adjustments. *J Exp Biol* 100: 5–22

Piiper J, Meyer M, Worth H, Willmer H (1977). Respiration and circulation during swimming activity in the dogfish *Scyliorhinus stellaris*. *Respir Physiol* 30: 221–239

Piiper J, Scheid P (1975). Transport efficacy of gills, lungs and skin: Theory and experimental data. *Respir Physiol* 23: 209–221

Piiper J, Scheid P (1982a). Models for comparative functional analysis of gas exchange organs in vertebrates. *J Appl Physiol* 53: 1321–1329

Piiper J, Scheid P (1982b). Physical principles of respiratory gas exchange in fish gills. In: Houlihan DF, Rankin JC, Shuttleworth TJ (eds) *Gills*. University Press, Cambridge, pp 45–61

Piiper J, Scheid P (1984). Model analysis of gas transfer in fish gills. In: Hoar WS, Randall DJ (eds) *Fish Physiology*. Academic Press, New York, London, vol XA, pp 229–262

Piiper J, Scheid P, Perry SF, Hughes GM (1986). Effective and morphometric O_2 diffusing capacity of the gills of the elasmobranch *Scyliorhinus stellaris*. *J Exp Biol* 123: 27–41

Piiper J, Schumann D (1967). Efficiency of O_2 exchange in the gills of the dogfish, *Scyliorhinus stellaris*. *Respir Physiol* 2: 135–148

Powell FL, Wagner PD (1982a). Measurement of continuous distributions of ventilation-perfusion in non-alveolar lungs. *Respir Physiol* 48: 219–232

Powell FL, Wagner PD (1982b). Ventilation-perfusion inequality in avian lungs. *Respir Physiol* 48: 233–241

Randall DJ, Holeton GF, Stevens ED (1967). The exchange of oxygen and carbon dioxide across the gill of rainbow trout. *J Exp Biol* 46: 339–348

Scheid P (1978). Analysis of gas exchange between air capillaries and blood capillaries in avian lungs. *Respir Physiol* 32: 27–49

Scheid P (1979). Mechanisms of gas exchange in bird lungs. *Rev Physiol Biochem Pharmacol* 86: 137–186

Scheid P, Hook C, Piiper J (1986). Models for analysis of countercurrent gas transfer in fish gills. *Respir Physiol* 64: 365–374

Scheid P, Piiper J (1970). Analysis of gas exchange in the avian lung: Theory and experiments in the domestic fowl. *Respir Physiol* 9: 246–262

Scheid P, Piiper J (1971a). Direct measurement of the pathway of respired gas in duck lungs. *Respir Physiol* 11: 308–314

Scheid P, Piiper J (1971b). Theoretical analysis of respiratory gas equilibration in water passing through fish gills. *Respir Physiol* 13: 305–318

Scheid P, Piiper J (1972). Cross-current gas exchange in avian lungs: Effects of reversed parabronchial air flow in ducks. *Respir Physiol* 16: 304–312

Schmidt-Nielsen K (1971). How birds breathe. *Sci Am* 225/6: 72–79

Shams H, Scheid P (1987) Respiration and blood gases in the duck exposed to normocapnic and hypercapnic hypoxia. Respir Physiol 67: 1–12

Tucker V (1968). Respiratory physiology of house sparrows in relation to high altitude flight. *J Exp Biol* 48: 55–66

14

Alterations in the Bimodal Gas Exchange of the African Catfish *Clarias lazera*

Amos Ar and David Zacks

To Hermann Rahn:
Who Predicted It All . . .

Introduction

The African catfish, an air-breathing teleost belonging to the order Cypriniformes family Claridae, is found in East Africa, including Sudan and Egypt, and in Israel. It inhabits freshwater pools, streams, lakes, and swamps where water exhibits variable levels of oxygen and carbon dioxide pressures (Pw_{O_2} and Pw_{CO_2}, respectively) both seasonally and daily. Pw_{O_2} may vary from oversaturation during daytimes with strong photosynthetic activity, down to very low values at nights in swamps with microbial activity (Abdel Magid 1971; Dejours 1975; Jordan 1976; Moussa 1956, 1957; Munshi 1961; Thomas et al. 1983).

Clarias lazera is a nocturnal predator. It is capable of leaving the water during humid nights to migrate through the wet grass from one water body to another (Babiker 1979; Bruton 1979; Jordan 1976; Moussa 1957; Singh and Hughes 1971). However, it is not capable of digging itself into the mud in order to survive the dry season (Babiker 1979; Bruton 1979). The ability of this fish to use aerobic respiration is due to special breathing organs, consisting of two suprabranchial chambers (SC) in the posterior dorsal part of the head, into which two pairs of tree-like organs (arborescent organs) protrude. These organs develop from the upper part of the second and fourth gill arches. The posterior pair is much larger than the anterior one. All are protected and separated from the underside by fan-like organs that are created from the tips of the gill arches. The walls of the chambers, the tree-like organs, and the "fans" are richly vascularized. The blood supply to all these organs is in parallel to that of the gills and is supplied by branches of the afferent branchial vessels. All efferent vessels merge into the dorsal aorta, and there are no shunts leading back to the heart (Moussa 1956; Munshi 1961; Satchell 1976; Singh et al. 1982).

The phenomenon of air breathing is well known in many species of tropical and subtropical freshwater fish, e.g., *Clarias batrachus* (Jordan 1976; Munshi 1961; Singh and Hughes 1971), *Clarias garpiensis* (Bruton 1979), *Saccobranchus fossilis* (Singh and Hughes 1971), *Symbranchus marmoratus* (Johansen 1966, 1968), *Amia calva* (Randall et al. 1981), *Protopterus annectens* (Babiker 1979; Johansen 1968), *Anabas testudineus* (Hughes and Singh 1970a, 1970b), *Arapaima gigas* (Johansen et al. 1978), and others.

Fish may be obligatory air breathers, such as *Electrophorus electricus* (Farber and Rahn 1970; Johansen 1968; Johansen et al. 1968), *Protopterus annectens* (Babiker 1979; Johansen and Lenfant 1968), and *Arapaima gigas* (Johansen et al. 1978; Stevens and Holeton 1978), which die if denied access to air even if there is no lack of oxygen in the water. Others are facultative air breathers, which use air mainly when facing a low oxygen pressure in water (Johansen 1968; Schmidt-Nielsen 1979), such as *Lepisosteus osseus* (Rahn et al. 1971), *Amia calva* (Randall et al. 1981), and *Channa punctata* (Hughes and Munshi 1973).

C. lazera is a facultative air breather. Its dependency on atmospheric oxygen is due to several factors, particularly a decrease of Pw_{O_2} and an increase in ambient temperature which induces increased activity (Abdel Magid 1971; Bruton 1979; Moussa 1957). According to Moussa (1957), large fish will die if denied air for more than 14–17 hr in aerated water. However, Abdel Magid and Babiker (1975) have found that if 50–80g fish are provided with aerated water at 25–26°C, they can survive for up to 14 days without access to air and obtain practically all the oxygen they need from the water. Singh and Hughes (1971) report that at 25°C small *Clarias batrachus* (76–105 g) obtain 58.4% of their oxygen by air breathing. In larger *C. batrachus* (150–200 g), Jordan (1976) found that the percentage of O_2 consumed from air increased from 4 to 75% when Pw_{O_2} was decreased from 150 to 25 Torr. In preliminary observations, we found that when *C. lazera* rises to the surface of the water, it exhales all the gas contained in the SC through the operculum openings by filling the SC with water. Upon reaching the surface, the fish opens its mouth, "swallows" fresh air in one gulp, and then sinks back into the bottom. It was also observed that when given a choice, *C. lazera* will prefer water of 27–29°C.

No definite results describing the alternatives in gas exchange of mature *C. lazera* (1.5–2.0 kg) at various Pw_{O_2} values with free access to normal atmosphere were found in the literature.

The aims of this study were to determine the rate of oxygen consumption, the rate of CO_2 production, and ventilation of this fish in water and air at preferred temperature and in different oxygen pressures in the water.

Materials and Methods

Mature *C. lazera* males of 1.65 kg ± 0.20 SD (56–71 cm long) were obtained from the Yigal Allon Kinneret Limnological Lab, Tiberias, Israel. Fish were kept separately in containers holding 70–90 l of aerated fresh water 30 cm deep at 27–30°C for several weeks before experiments. The water level was checked, and the fish were fed ad lib standard rat pellets and fly larvae, twice a week. Experiments were carried out during daytime hours. Before the experiments, the fish were starved for at least 24 hr. The fish were weighed and transferred into an "open-system"-type metabolic chamber. The metabolic chamber consisted of a large "perspex" compartment, $60 \times 13 \times 13$ cm (10.14 l), on the top of which an additional small compartment, $13 \times 12 \times 9$ cm (1.40 l), was mounted such that both compartments formed an L-shaped metabolic chamber. The large compartment was filled with water which was pumped through it by an Eheim (Model 6) pump from a 100-l reservoir. The small compartment was kept air-filled. Water flow was regulated at a rate of 1.25 to 1.50 $l \cdot min^{-1}$. The water temperature was kept at 28.5°C ± 0.5 with a thermostated 100-W heater (Joger type R). The Pw_{O_2} level was obtained by bubbling air, nitrogen, or oxygen at different rates into the reservoir. Pw_{CO_2} did not exceed 1 Torr. The fish inhaled air from the small closed compartment; after every air gulp, the compartment was opened and aerated and its water floor was brought to the previous level. The apparatus was positioned at such an angle that before the air exhaled from the operculi of the fish reached the small compartment, it was trapped under the ceiling of the large compartment and accumulated at the large compartment's rear end, from which it was collected in its entirety via a special capped opening into a 100-ml syringe.

The fish were equilibrated to experimental conditions for 2 hr before experiments began. Each experiment lasted at least 4 hr. The Pw_{O_2} of the incoming water was measured by sampling the water with a 5-ml syringe every few minutes. Water samples were analyzed with a Radiometer PHM72 MK2 analyzer connected to a Radiometer E5047 oxygen electrode mounted in a type D616 glass thermostated cell. The electrode was thermostated with a Haake Model FS water thermostat at 37°C ± 0.1 to ensure fast response. Ascending time was noted during the experiments in order to calculate the time intervals between gulps. Water ventilation frequency, fW, was measured from mouth or operculum movements immediately after descent, occasionally during bottom stay time, and in many cases just before the next ascent. Gaseous O_2 and CO_2 concentrations were measured with an Applied Electrochemistry S-3A oxygen analyzer. The O_2 concentration was read directly after drying the injected gas sample by passing it through a Drierite column (W.A. Hammond) made out of a 1-ml tuberculin syringe. The CO_2 concentrations were indirectly calculated from differences between the first O_2 reading and a second reading taken after the gas had been passed through a similar Ascarite column (A.H. Thomas) to absorb CO_2 and water vapor.

Aerial O_2 consumption rates $\dot{M}A_{O_2}$, and rates of CO_2 released to air $\dot{M}A_{CO_2}$ were calculated from the measured volumes of exhaled gas, the O_2 and CO_2 concentrations of the exhaled air, and the time intervals between gulps, correcting for the R, temperature, and vapor effects on volume.

For obtaining data concerning the change in PA_{O_2} and PA_{CO_2} in the SC, two other fish were used. At any time needed, 2-ml gas samples from the SC could be obtained directly via a PE 50 polyethylene tube, using a syringe needle that was introduced permanently into the upper distal part of the SC under MS 222 (100 mg·l^{-1}) anesthesia. The fish were allowed to recover for at least 24 hr before experiments began. The tubes were long enough to allow free-swimming behavior.

Results

The rates of oxygen consumption in air, $\dot{M}A_{O_2}$ and in water $\dot{M}w_{O_2}$ of four mature African catfish, *C. lazera,* in different Pw_{O_2} ranges are presented in Table 14.1. Table 14.1 values show that in Pw_{O_2} near saturation and in oversaturation the need for aerial oxygen is marginal. Fish no. 4 in Table 14.1 showed a high $\dot{M}A_{O_2}$ even in high Pw_{O_2}. This fish was not preequilibrated long enough before the experiments, appeared to be restless throughout the experiments, and yielded only partial results. We did not use the data concerning $\dot{M}A_{O_2}$ to $\dot{M}w_{O_2}$ ratios in the different Pw_{O_2}'s of this fish for further calculations. $\dot{M}A_{O_2}$ increased as Pw_{O_2} decreased. It reached about 50% of the total consumed at about 50% water O_2 saturation. At the same time $\dot{M}w_{O_2}$ decreased. It seems, however, that

TABLE 14.1. The rates of gas exchange in air and water at different oxygen pressures Pw_{O_2}, their totals, the ratios of gas exchanges in air to totals, and the respiratory coefficients R in air and water of *Clarias lazera*.*

Fish no. Mass(kg)	Pw_{O_2} (Torr)	$\dot{M}A_{O_2}$ [ml STPD·(kg·hr)⁻¹]	$\dot{M}w_{O_2}$ [ml STPD·(kg·hr)⁻¹]	$\dot{M}T_{O_2}$	$\dot{M}A_{CO_2}$ [ml STPD·(kg·hr)⁻¹]	$\dot{M}w_{CO_2}$ [ml STPD·(kg·hr)⁻¹]	$\dot{M}T_{CO_2}$	$\dfrac{\dot{M}A_{O_2}}{\dot{M}T_{O_2}}$	$\dfrac{\dot{M}A_{CO_2}}{\dot{M}T_{O_2}}$	R in air	R in water
(1)	117	2.3	49.6± 6.1	51.9	0.08	44.0	44.1	0.044	0.002	0.03	0.9
	100	4.1± 0.8	47.0± 5.3	51.7	0.14±0.04	43.8	43.9	0.099	0.003	0.03	0.9
	94	19.1± 3.6	41.1± 5.8	60.2	0.58±1.10	50.7	51.2	0.317	0.010	0.03	1.2
1.84	78	15.9± 2.8	41.4± 5.9	57.3	1.12±0.17	47.6	48.7	0.277	0.023	0.07	1.1
	74	20.6± 8.5	29.3± 4.3	49.9	0.70±0.19	41.7	42.4	0.413	0.017	0.03	1.4
	65	29.1± 4.0	32.8± 3.1	61.9	1.91±0.59	50.7	52.6	0.470	0.036	0.06	1.5
Mean±SD				55.5±5.0			47.2±4.2			0.04±0.02	1.2±0.3
(2)	150	5.3± 1.3	65.6± 4.7	70.9	0.46±0.02	59.7	60.2	0.075	0.008	0.09	0.9
	131	4.9	57.1± 5.5	62.0	0.43	52.3	52.7	0.079	0.008	0.09	0.9
	122	5.0	43.5± 1.0	48.0				0.104			
1.96	105	20.4±10.6	32.6± 0.8	53.0	0.57±0.34	44.5	45.05	0.385	0.013	0.03	1.4
	85	33.8±19.0	25.7± 5.5	59.5	3.70±3.10	46.9	50.6	0.568	0.073	0.11	1.8
	78	31.8± 8.9	24.2± 2.4	55.8	2.56±0.41	44.8	47.4	0.566	0.054	0.08	1.9
Mean±(SD)				58.5±7.9			49.5±6.7			0.08±0.03	1.4±0.5

(3)	198	1.3	56.7± 5.7	58.0	0.03	49.3	49.3	0.022	0.001	0.02	0.9
	176	2.4	55.4± 7.1	57.8	0.01	49.1	49.1	0.041	0.0002	0.004	0.9
	135	7.5± 3.5	43.3±12.0	50.8	0.34±0.09	42.9	43.2	0.148	0.008	0.04	1.0
1.50	128	13.8± 6.9	51.2± 9.4	65.0	0.42±0.16	54.8	55.2	0.212	0.008	0.03	1.1
	115	39.6±11.9	31.3± 3.3	70.9	4.49±1.00	55.8	60.3	0.558	0.074	0.11	1.8
	91	42.4± 9.5	23.5± 2.9	65.9	4.57±0.82	51.4	56.0	0.643	0.082	0.11	2.2
	88	28.9± 6.9	25.7± 6.0	54.6	2.94±0.67	43.4	46.4	0.529	0.063	0.10	1.7
Mean±SD				63.8±11.5			54.2±9.8			0.7±0.04	1.7±1.0
(4)	145	17.5± 1.5	56.8± 4.3	74.3	3.11±1.19	60.0	63.1	0.235	0.049	0.18	1.0
1.40	132	33.9±11.2	27.0± 5.6	60.9	3.29±0.54	48.5	51.8	0.556	0.063	0.10	1.8
	123	35.6±11.1	28.3± 5.3	63.9	4.65±0.25	49.6	54.3	0.557	0.086	0.13	1.7
Mean±SD				66.4±7.0			56.4±6.0			0.13±0.04	1.5±0.4

* $\dot{M}_{A_{O_2}}$, $\dot{M}_{A_{CO_2}}$, $\dot{M}_{W_{CO_2}}$, $\dot{M}_{W_{O_2}}$, $\dot{M}_{T_{O_2}}$, and $\dot{M}_{T_{CO_2}}$ are the rates of gas exchange in air and water and the totals for O_2, and CO_2 respectively. Experimental temp. = 28.5°C.

the total, $\dot{M}_{TO_2} = (\dot{M}_{AO_2} + \dot{M}_{WO_2})$ in resting fish, was not affected, as shown in Table 14.2.

From measurements and calculations presented in Table 14.1, it can be seen that most of the total CO_2 extracted ($\dot{M}_{TCO_2}$), is exchanged through the water ($\dot{M}_{WO_2}$). The percentage of CO_2 exchanged into air ($\dot{M}_{ACO_2}$) increases as P_{WO_2} decreases (Fig. 14.1), although the fish stays under water between gulps, for shorter periods (Table 14.3). A comparison of the relative rates of exchange of O_2 and CO_2 in air and water indicates that the main sites of exchange for each of these two gases must be differently localized. The main organ for the CO_2 exchange is apparently not coupled with the SC.

The simultaneous values of P_{AO_2} and P_{ACO_2} in the SC as measured directly in normoxic and hypoxic water in expired and sampled gas are shown in Fig. 14.2. R values for air and water are given in Table 14.1. Fig. 14.2 shows that in normoxic water (NW) the P_{ACO_2}/P_{AO_2} values in air fall on low R lines (R = 0.07 ± 0.03 SD), while in hypoxic water (HW) the R values are higher (0.50 ± 0.19 SD). Both the values in Fig. 14.2 and the calculations from Table 14.3 show that P_{AO_2} may drop down to 12 Torr. Thus, the gaseous O_2 extraction coefficient may reach 92%, and is on the average 82.9% ± 5.5 SD. The rates of the P_{AO_2} drop in the SC with time are shown in Fig. 14.3. It is clear from Fig. 14.3 that the $\dot{M}_{AO_2}$ in the SC between gulps is not constant. It declines with time, and this is shown in Fig. 14.6.

Table 14.3 summarizes the rates of air and water respiration. Fig. 14.4 describes the ascending rate (aerial respiration rate, fA), of *C. lazera* as a function of P_{WO_2}. It can be seen that fA increases as P_{WO_2} decreases, but stays variable in all P_{WO_2} values. The coefficient of variation of the fA ranges between 20 and 50%. The regression equation describing the relationship between fA and $\dot{P}_{WO_2}$ is fA = $43.81 \cdot e^{-0.025 \cdot PwO_2}$ (r = −0.986).

The respiratory rate in water, fW, measured from the operculum and mouth movements increased gradually with elapsed time between gulps

TABLE 14.2. The sums of oxygen consumption rates [ml STPD·(kg·hr)$^{-1}$] from water ($\dot{M}_{WO_2}$) and from air ($\dot{M}_{AO_2}$), and their proportions in percentages, in different oxygen pressures in the water (P_{WO_2}).*

P_{WO_2} range (Torr)	60–89	90–119	120–149	150 and up
(No. of fish)	(3)	(3)	(2)	(2)
($\dot{M}_{WO_2}$ + $\dot{M}_{AO_2}$) ± SD	60.9 ± 12.2	58.9 ± 8.1	56.4 ± 8.3	62.2 ± 7.5
$\dot{M}_{WO_2}$ (%)	48	66	86	95.4
$\dot{M}_{AO_2}$ (%)	52	34	14	4.6

* Water temperature = 28.5°C ± 0.5.

TABLE 14.3. The rate of aerial ventilation ($\dot{V}_A$), aerial respiration frequency (fA), single expired gulp volume, oxygen pressure in expired gas ($P_{A_{O_2}}$), aerial convection requirement, and O_2 extraction coefficient of *Clarias lazera* in water of different O_2 pressures (Pw_{O_2}) at 28.5°C.

Fish No. Mass (kg)	Pw_{O_2} (Torr)	Gulp vol (ml)	fA (1 hr^{-1})	$\dot{V}_A$ (ml·hr^{-1})	Final $P_{A_{O_2}}$ (Torr)	Convect. requir. (ml·ml^{-1})	Extract. coeff. (%)
(1)	117	84	0.4	29.4	23.3	12.8	85.3
	100	81	0.6±0.0	50.2	20.8	10.7	86.8
	94	79	3.0±0.6	237.0	31.7	12.4	80.0
1.84	78	83	2.3±0.6	190.9	17.7	12.0	88.8
	74	78	4.3±2.3	335.4	26.6	16.3	83.2
	65	76	4.8±1.5	364.8	22.6	12.5	85.7
Mean±SD		80±3			23.8±4.9	12.8±1.9	84.9±3.0
(2)	150	78	0.7	51.5	12.1	9.7	92.3
	131	71	0.6	45.4	17.7	9.3	88.8
	122	76	0.8	60.0	6.0	12.0	96.2
1.96	105	73	4.5±2.2	328.5	19.0	16.1	88.0
	85	69	6.5±3.5	448.5	23.7	13.3	85.0
	78	68	6.5±3.3	442.0	24.7	14.0	84.4
Mean±SD		73±4			17.2±7.1	12.4±2.6	89.1±4.5
(3)	198	70	0.3	21.0	67.8	16.1	57.1
	176	73	0.4	29.2	31.9	12.2	79.8
	135	69	1.4±0.5	96.6	42.4	12.9	73.2
	128	70	1.5±1.0	105.0	49.3	17.2	68.8
1.50	115	67	6.4±1.3	428.8	33.4	10.8	78.9
	91	68	5.7±2.3	387.6	21.8	9.1	86.2
	88	65	4.7±1.5	305.5	18.6	10.6	88.2
	73	63	11.3±4.3	711.9	37.7	10.1	76.1
Mean±SD		68±3			37.9±15.7	12.4±2.9	76.0±9.9
(4)	145	58	6.9±0.7	400.2	21.9	22.9	86.15
1.40	132	56	8.2±1.7	459.2	27.4	13.5	82.7
	123	51	9.2±2.2	494.7	37.5	13.9	76.3

(Fig. 14.5). Immediately after descent it was fW = 24 min^{-1} ± 2 SD and increased to fW = 44 min^{-1} ± 4 SD just before the next ascent. We observed, although we could not quantify, a gradual increase in the operculum opening and mouth opening amplitudes in the same time interval.

Fig. 14.6 describes the instantaneous values of $\dot{M}w_{O_2}$, derived from measurements made for various time intervals between gulps. By using the natural time of gulp variation, it was possible to calculate instantaneous $\dot{M}a_{O_2}$ values as a function of elapsed time between gulps that were all finally superimposed on the same time scale. It can be seen that

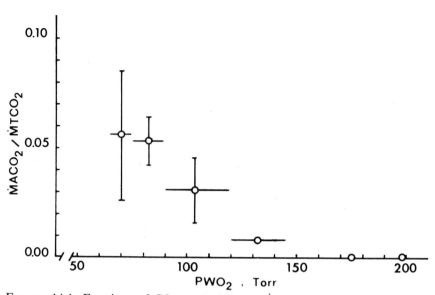

FIGURE 14.1. Fractions of CO_2 expired to air ($\dot{M}A_{CO_2}$) from total expired CO_2 ($\dot{M}T_{CO_2}$) of *C. lazera* at 28.5°C, at different O_2 pressures in water (Pw_{O_2}). Vertical bars: SE. Horizontal lines: Range.

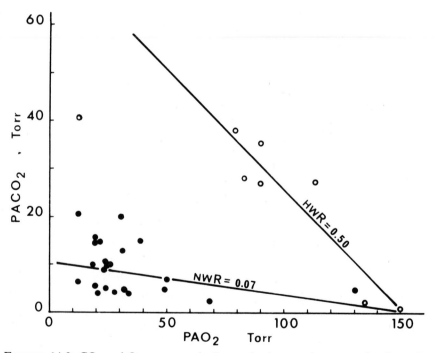

FIGURE 14.2. CO_2 and O_2 pressures in the expired gas and gas samples from the suprabranchial chambers (SC) of *C. lazera* (PA_{CO_2} and PA_{O_2}, respectively). Closed circles: Normoxic water. Open circles: Hypoxic water. NWR and HWR are the respiratory quotients for normoxic and hypoxic water, respectively. Data points for HWR represent gas samples from SC only.

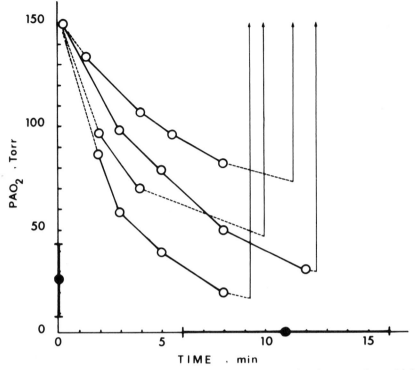

FIGURE 14.3. Examples of the change in O_2 pressure in the suprabranchial chambers, (P_{AO_2}) of a 3-kg *C. lazera* in normoxic water with time. Arrows indicate ascent times. The average final P_{AO_2} in this fish ($\pm$ 2 SD) is indicated on the ordinate; the mean ascent time ($\pm$ 2 SD) is indicated on the abscissa (closed circles) ($n = 30$).

instantaneous $\dot{M}w_{O_2}$ increases gradually between gulps, while $\dot{M}A_{O_2}$ decreases at the same time interval.

Data given in Table 14.3 show that air ventilation ($\dot{V}_A$) increases as Pw_{O_2} decreases (Fig. 14.4). The regression equation describing the relationship between $\dot{V}_A$ and Pw_{O_2} is $\dot{V}_A = 2369 \cdot e^{-0.026 \cdot Pw_{O_2}}$ ($r = 0.845$). However, gulp volume is unchanged and is on the average 70.7 ml $\pm$ 8.4 SD. The convection requirement stays unaffected in all Pw_{O_2}'s at 12.5 ml·ml^{-1} $\pm$ 2.4 SD (Table 14.3).

Fig. 14.7 describes the relationship between the time interval between gulps and average rate of oxygen consumed from the gas in the SC during this interval. The figure expresses the time interval as fA (hr^{-1}). There is a linear relationship between the reciprocal value of each time interval and the mean specific O_2 consumption from the gulp volume to which a particular time interval belongs.

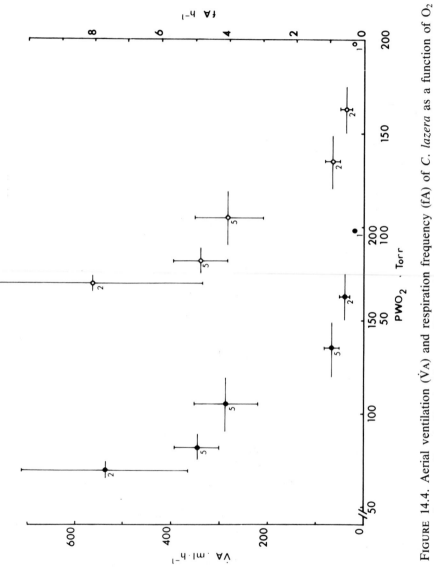

FIGURE 14.4. Aerial ventilation ($\dot{V}_A$) and respiration frequency (fA) of *C. lazera* as a function of O_2 pressure in water (Pw_{O_2}). Vertical bars: ± SD. Horizontal bars: Pw_{O_2} range.

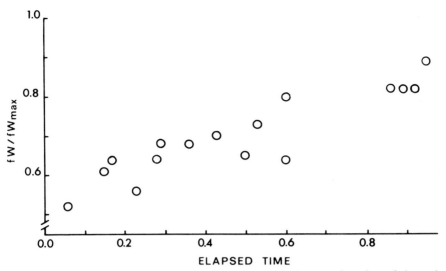

FIGURE 14.5. Changes in water respiration frequency fW as a function of elapsed time between gulps. Both scales are normalized to the fraction of the total.

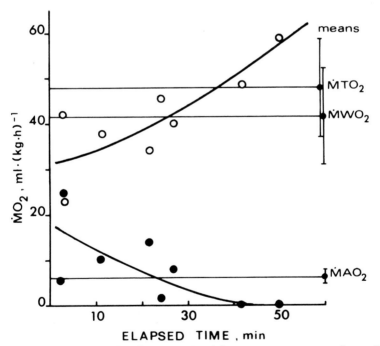

FIGURE 14.6. An example (fish no. 3) of instantaneous rates of specific O_2 consumption ($\dot{M}O_2$) in water and air during elapsed time between gulps in normoxia; for details, see text. Open circles: Specific O_2 consumption from water ($\dot{M}W_{O_2}$). Closed circles: The same from air ($\dot{M}A_{O_2}$). Long-term averages of total O_2 consumption ($\dot{M}T_{O_2}$), $\dot{M}W_{O_2}$, and $\dot{M}A_{O_2}$ are indicated with their $\pm$ SD. Lines are fitted by eye.

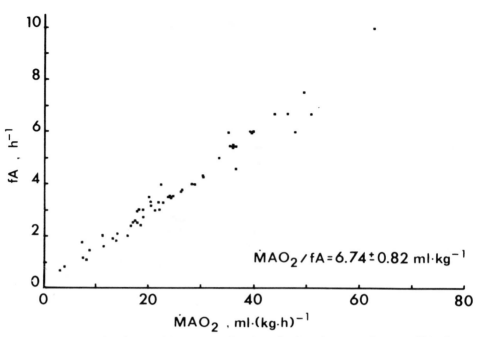

FIGURE 14.7. Time interval between gulps described as the rate of ascent (fA) of *C. lazera* (fish nos. 1, 2, and 3) as a function of their mean specific rate of O_2 consumption from the air enclosed in the suprabranchial chambers during the same time interval, $\dot{M}_{AO_2}$. Data include values in all P_{WO_2}'s measured.

Discussion

While the lungfish represents a special evolutionary trend, teleosts that acquired the ability to utilize aerobic oxygen probably evolved independently more than once during evolution, and represent a secondary adaptation to life in hypoxic water (Liem 1980). This is reflected in the various aerobic respiratory organs that different teleosts have developed (Abdel Magid and Babiker 1975; Carter and Beadle 1931; Johansen 1968; Rahn and Howell 1976; Satchell 1976; Schmidt-Nielsen, 1979).

It seems that air-breathing teleosts are not as dependent on air respiration as are other air-breathing fish. This can be demonstrated by comparing the proportions of oxygen obtained from air in normoxic (NW) and hypoxic (HW) water amongst mature fish, in a comparable range of temperature and P_{WO_2}, to those to which we subjected *C. lazera* (Fig. 14.8). The figure suggests that air breathing in teleosts represents mainly an emergency response to temporary hypoxic conditions, evolving from the behavior of some water-breathing teleosts which surface in hypoxia and utilize aquatic surface respiration (Burggren 1982; Kramer 1983).

The demands for oxygen of *C. lazera* are not essentially different from those of other fish of similar masses in similar conditions of temperature and Pw_{O_2}, e.g., *Amia calva* (Randall et al. 1981) and *Arapaima gigas* (Stevens and Holeton 1978). Even among the teleosts (Fig. 14.8) *C. lazera* cannot be considered as a very efficient air breather, and in HW (Pw_{O_2} = 65–80 Torr), it obtains only about 50% of its O_2 consumption from air. Still it can survive for a few days on air breathing during migration from one pond to another (Bruton 1979; personal observations). Some preliminary observations show that mature fish can adapt to water breathing in half-saturated water and relatively high temperature (27°C), when denied

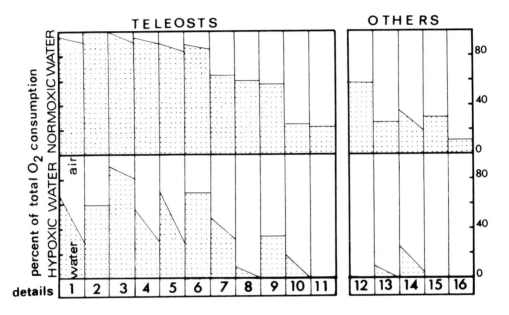

FIGURE 14.8. O_2 consumption rate in water and air of air-breathing fish. Scale units are fractions of the total. The diagonal line in each column represents the range (if given). 1. *Clarias lazera*, 1.7 kg at 28.5°C (present work). 2. *Saccobranchus fossilis*, 50–60 g at 25°C (Hughes and Singh 1971). 3. *Clarias batrachus*, 150–210 g at 25°C (Jordan 1976). 4. *Piabucina festae*, 20–80 g at 25°C (Graham et al 1977). 5. *Hoplosternum thoracatum*, 5–10 g at 24°C (Gee and Graham 1978). 6. *Anguilla anguilla*, 400–600 g at 15°C (Berg and Steen 1965). 7. *Gymnotus carapo*, 30–75 g at 30°C (Liem et al. 1984). 8. *Calamoichthys calabaricus*, 20–30 g at 27° (Sacca and Burggren 1982). 9. *Trichogaster trichogaster*, 8 g at 27°C (Burggren 1979). 10. *Arapaima gigas*, 2–3 kg at 28–29°C (Stevens and Holeton 1978). 11. *Electrophorus electricus*, 2.76 kg at 26°C (Farber and Rahn 1970). 12. *Lepisosteus oculatus*, 600–1400 g at 20°C (Smatresk and Cameron 1982). 13. *Lepisosteus osseus*, 300–900 g at 22°C (Rahn et al. 1971). 14. *Amia calva*, 750–1650 g at 27°C (Johansen et al. 1970). 15. *Lepidosiren paradoxa*, 10–20 g at 20°C (Johansen and Lenfant 1967). 16. *Protopterus aethiopicus*, 500 g and up at 20°C (Lenfant and Johansen 1968).

access to air (Zacks and Becker, unpublished). Abdel Magid (1971) and Abdel Magid and Babiker (1975) showed the same in young *C. lazera* which were forced to stay submerged.

The $\dot{M}T_{O_2}$ of *C. lazera*, in the range of Pw_{O_2} from 65 to 200 Torr, is nearly constant, and only the proportion between $\dot{M}A_{O_2}$ and $\dot{M}w_{O_2}$ changes (Tables 14.1 and 14.2). Similar results were obtained by Berg and Steen (1965) for *Anguilla anguilla*, by Farber and Rahn (1971) for *Electrophorus electricus*, by Stevens and Holeton (1978) for *Arapaima gigas*, by Burggren (1979) for *Trichogaster trichopterus*, and by Yu and Woo (1985) for *Channa maculata*. Unlike those fish, both Hughes and Singh (1971) and Graham et al (1977) showed for *Saccobranchus fossilis* and *Piabucina festae*, respectively, that $\dot{M}T_{O_2}$ drops with Pw_{O_2} to a new, lower, stabilized level.

The $\dot{M}T_{CO_2}$ was not measured directly in this study. It was calculated together with $\dot{M}w_{CO_2}$ (Table 1) using the assumption that the overall RQ of the fish is 0.85 (Rahn and Howell 1976; Singh 1976; Becker, personal communication). Fig. 14.1 shows that the fraction of $\dot{M}A_{CO_2}$, from the total $\dot{M}T_{CO_2}$, is much higher in HW compared with NW. This increase in $\dot{M}A_{CO_2}$ is more than in proportion to the increase in $\dot{M}A_{O_2}$ in HW, as may be judged from the changes in R values in water and air (Table 14.1, Fig. 14.2). Similar results are presented by Singh (1976) for *Anabas testudineus*, *Saccobranchus fossilis*, and *Clarias batrachus*. The results of Yu and Woo (1985) show no change in aerial R for *Channa maculata*. Our results indicate a relative shift in the blood flow from the gills (and skin) to the arborescent organs (and the walls of the SC). It seems that only in the genus *Clarias* and in *Saccobranchus fossilis*, the blood supplies to the aerial and to the water respiratory organs are completely in parallel (Moussa 1956; Satchell 1976; Singh and Hughes 1971); and therefore, blood may be shunted effectively from one organ to the other in accordance with the availability of oxygen in the respiratory media. It would be interesting to test if such blood flow changes do occur. The supporting evidence we have is that the CO_2 fraction in expired gas tends to increase in hypoxic water, at constant low Pw_{CO_2} (Fig. 14.1), and that this is done in spite of the fact that the rates of both air and water ventilations increase in HW (Figs. 14.4 and 14.5). These changes indicate an increase in P_{CO_2} in the blood supplying the aerial respiratory organs. Moreover, $\dot{M}w_{CO_2}$ does not increase in HW and may even decrease (Table 14.1). Thus, CO_2 is not exchanged efficiently through the gills in HW, indicating a relatively low blood flow there.

The increase in CO_2 pressure in the blood may affect O_2 exchange through the presence of the Haldane and Root effects. "Some" Root effect was found in *Clarias batrachus* (Singh 1976), but gas-blood properties are not available for *C. lazera*.

Using the variation of the intervals between two subsequent gulps in different Pw_{O_2}'s and the analysis of the expired gas, we have calculated mean $\dot{M}_{A_{O_2}}$ for different time intervals between gulps. This is shown in Figure 14.7, where the time intervals between gulps are expressed as fA. All experimental points for all fish in all $P_{W_{O_2}}$'s fall on a common line. The reciprocal value of the slope has units of volume of oxygen per unit of mass. Thus, a fixed volume of 6.74 ml O_2/kg body mass is the amount of O_2 extracted from an air gulp in resting fish independently of $P_{W_{O_2}}$. However, this volume is consumed faster in low $P_{W_{O_2}}$ reaching the same extraction coefficient of 82.9% in the presence of relatively high $P_{A_{CO_2}}$ and presumably high blood P_{CO_2}. Thus, gas-blood properties must favor O_2 binding under such conditions. This is in contrast to the situation in *Channa maculata*, where the air-breathing organ has a limited rate of O_2 absorption in hypoxic water and a fixed aerial R value (Yu and Woo 1985). From comparing the equations that describe both fA and $\dot{V}_A$ as a function of $P_{W_{O_2}}$ (Fig. 14.4), it is evident that the gulp volume of 71 ml is unchanged with $P_{W_{O_2}}$, and thus only the rate of emergence contributes to the increase in aerial respiration. Since in all conditions *C. lazera* have the same amount of O_2 extracted, the same gulp volume, the same O_2 extraction coefficient, and the same convection requirement (Table 14.3), the mechanism of air breathing seems to be of fixed capabilities in terms of oxygen delivery.

An interesting feature of *C. lazera* is the relative changes that occur between air and water respiration in the time interval between two gulps. Changes in $P_{A_{O_2}}$ and $P_{A_{CO_2}}$ in the SC with time are given by Singh and Hughes (1973) for *Anabas testudineus*. A continuous but not linear decline in $P_{A_{O_2}}$ in the SC between gulps is shown in Fig. 14.3, while at the same time fW increases (Fig. 14.5), indicating possible changes in the gas-exchange proportion between water and air. By using the average of $\dot{M}_{T_{O_2}}$ and the average of $\dot{M}_{A_{O_2}}$ and using instantaneous measurements of $\dot{M}_{W_{O_2}}$ at different times after gulping (corrected for time lag of the measurement), and then superimposing the results from different gulps on the same time scale, we were able to construct the course of changes in $\dot{M}_{W_{O_2}}$ and $\dot{M}_{A_{O_2}}$ during one interval between gulps. One such example is shown in Fig. 14.6. The two continuous complementary changes in $\dot{M}_{O_2}$ indicate that in each time interval between two successive gulps the instantaneous rate of O_2 extracted from the SC declines, and the rate of O_2 extracted from water increases with time to preserve the total O_2 consumed at all times.

Acknowledgments. We thank Prof. Lev Fishelson from our Department and Dr. Klaus Becker from Göttingen University for helpful discussions and Mr. Nissim Sharon for his assistance in handling the fish. The Yigal Allon Kinneret Limnological Laboratory, Tiberias, kindly provided the fish. Special thanks go to Ms. Ann Belinsky, for her devoted assistance throughout this work.

REFERENCES

Abdel Magid AM (1971). The ability of *Clarias lazera* (Pisces) to survive without air breathing. *J Zool Lond* 163: 63–72

Abdel Magid AM, Babiker MM (1975). Oxygen consumption and respiratory behaviour of three Nile fishes *Hydrobiologia* 46: 359–367

Babiker MM (1979). Respiratory behaviour, oxygen consumption and relative dependence on aerial respiration in the African lungfish (*Protopterus annectens*, Owen) and an air breathing teleost (*Clarias lazera*, C.). *Hydrobiologia* 65: 177–187

Berg T, Steen JB (1965). Physiological mechanisms for aerial respiration in the eel. *Comp Biochem Physiol* 15: 469–484

Bruton MN (1979). The survival of habitat desiccation by air breathing clariid catfishes. *Env Biol Fish* 4: 273–280

Burggren WW (1979). Bimodal gas exchange during variation in environmental oxygen and carbon dioxide in the air breathing fish *Trichogaster trichopterus*. *J Exp Biol* 82: 197–213

Burggren WW (1982). "Air gulping" improves blood oxygen transport during aquatic hypoxia in the goldfish *Carassius auratus*. *Physiol Zool* 55: 327–334

Carter GS, Beadle LC (1931). The fauna of the Paraguayan chaco. *J Zool Lond* 37: 327–368

Dejours P (1975). *Principles of Comparative Respiratory Physiology*. North-Holland Publishing Co., Amsterdam, 253 pp

Farber J, Rahn H (1970). Gas exchange between air and water and the ventilation pattern in the electric eel. *Respir Physiol* 9: 151–161

Gee JH, Graham JB (1978). Respiratory and hydrostatic functions of the intestine of the catfishes *Hoplosternum thoracatum* and *Brochis splendens* (Callichthyidae). *J Exp Biol* 74: 1–16

Graham JB, Kramer DL, Pineda E. (1977). Respiration of the air breathing fish *Piabucina festae*. *J Comp Physiol* 122: 295–310

Hughes GM, Singh BN (1970a). Respiration in an air-breathing fish, the climbing perch, *Anabas testudineus*. I. Oxygen uptake and carbon dioxide release into air and water. *J Exp Biol* 53: 265–280

Hughes GM, Singh BN (1970b). Respiration in the air-breathing fish, the climbing perch, *Anabas testudineus*. II. Respiratory patterns and the control of breathing. *J Exp Biol* 53: 281–298

Hughes GM, Singh BN (1971). Gas exchange with air and water in an air breathing catfish *Saccobranchus* (*Heteropneustes*) *fossilis*. *J Exp Biol* 55: 667–682

Hughes GM, Munshi JSD (1973). Nature of the air-breathing organ of the Indian fishes: *Channa, Amphipnous, Clarias* and *Saccobranchus* as shown by electron microscopy. *J Zool Lond* 170: 245–270

Johansen K (1966). Air breathing in the teleost *Symbranchus marmoratus*. *Comp Biochem Physiol* 18: 383–395

Johansen K (1968). Air breathing fishes. *Sci Am* 219: 102–111

Johansen K, Hanson D, Lenfant C (1970). Respiration in a primitive air breather *Amia calva*. *Respir Physiol* 9: 162–174

Johansen K, Lenfant C (1967). Respiratory function in the South American lungfish, *Lepidosiren paradoxa* (*Fitz.*) *J Exp Biol* 46: 205–218;

Johansen K, Lenfant C (1968). Respiration in the African lungfish *Protopterus aethiopicus. II. Control of Breathing J Exp Biol* 49: 453–468

Johansen K, Lenfant C, Schmidt-Nielsen K, Peterson K (1968). Gas exchange and the control of breathing in the electric eel *Electrophorus electricus. Z Vergl Physiol* 61: 137–163

Johansen K, Mangum CP, Weber RE (1978). Reduced blood O_2 affinity associated with air breathing in osteoglossid fishes. *Can J Zool* 56: 891–897

Jordan J (1976). The influence of body weight on gas exchange in the air breathing fish *Clarias batrachus. Comp Biochem Physiol* 53A: 305–310

Kramer DL (1983). The evolutionary ecology of respiration mode in fishes: An analysis based on the costs of breathing. *Env Biol Fishes* 9: 145–158

Lenfant C, Johansen K (1968). Respiration in the African lungfish *Protopterus aethiopicus.* I. Respiratory properties of blood and normal patterns of breathing and gas exchange. *J Exp Biol* 49: 437–452

Liem KF (1980). Air ventilation in advanced teleosts: Biochemical and evolutionary aspects. In: Ali MA (ed) *Environmental Physiology of Fishes.* Plenum Press, New York, pp 57–91

Liem KF, Eclancher B, Fink WL (1984). Aerial respiration in the banded knife fish *Gymnotus carapo* (Teleostei: Gymnotoidei). *Physiol Zool* 51: 185–195

Moussa TA (1956). Morphology of the accessory air breathing organs of the teleost *Clarias lazera* (C and V). *J Morphology* 98: 125–160

Moussa TA (1957). Physiology of the accessory respiratory organs of the teleost, *Clarias lazera* (C and V). *J Exp Zool* 136: 419–454

Munshi JSD (1961). The accessory respiratory organs of *Clarias batrachus. J Morph* 109: 115–140

Rahn H, Howell BJ (1976). Bimodal gas exchange. In: Hughes GM (ed) *Respiration of Amphibious Vertebrates.* Academic Press, New York, pp 271–285

Rahn H, Rahn KB, Howell BJ, Ganz C, Tenney SM (1971). Air breathing of the garfish (*Lepisosteus osseus*) *Respir Physiol* 11: 285–307

Randall DJ, Cameron JN, Daxboeck C, Smatresk NJ (1981). Aspects of bimodal gas exchange in the bowfin *Amia calva* L. (Actinopterygii: Amiiformes). *Respir Physiol* 43: 339–348

Sacca R, Burggren WW (1982). Oxygen uptake in air and water in the air-breathing reedfish *Calamoichthys calabaricus.* Role of skin, gills and lungs. *J Exp Biol* 97: 179–186

Satchell GH (1976). The circulatory system of air-breathing fish. In: Hughes GM (ed) *Respiration of Amphibious Vertebrates.* Academic Press, New York, pp 105–124

Schmidt-Nielsen K (1979). Air-breathing fish. In: *Animal Physiology: Adaptation and Environment,* 2nd ed. Cambridge University Press, Cambridge, pp 37–43

Singh BN (1976). Balance between aquatic and aerial respiration. In: Hughes GM (ed) *Respiration of Amphibious Vertebrates.* Academic Press, New York, pp 125–164

Singh BN, Hughes GM (1971). Respiration of an air-breathing catfish *Clarias batrachus* (Linn.). *J Exp Biol* 55: 421–434

Singh BN, Hughes GM (1973). Cardiac and respiratory responses in the climbing perch *Anabas testudineus. J Comp Physiol* 84: 205–226

Singh BR, Mishra AP, Sheel M, Singh I (1982). Development of the air-breathing organ in the cat fish, *Clarias batrachus* (Linn.). *Zool Anz Jenna* 208: 100–111

Smatresk NJ, Cameron JN (1982). Respiration and acid-base physiology of the spotted gar, a bimodal breather. I. Normal values, and the response to severe hypoxia. *J Exp Biol* 96: 263–280

Stevens ED, Holeton GF (1978). The partitioning of oxygen uptake from air and from water by the large obligate air-breathing teleost pirarucu (*Arapaima gigas*). *Can J Zool* 56: 974–976

Thomas S, Fievet B, Barthélémy L, Peyraud C (1983). Comparison of the effects of exogenous and endogenous hypercapnia on ventilation and oxygen uptake in the rainbow trout (*Salmo gairdneri* R.). *J Comp Physiol* 151: 185–190

Yu KL, Woo NYS (1985). Effects of ambient oxygen tension and temperature on the bimodal respiration of an air-breathing teleost *Channa maculata*. *Physiol Zool* 58: 181–189

15

Homeostasis: Embracing Negative Feedback Enhanced and Sustained by Positive Feedback

H.T. HAMMEL

Introduction

Comparative physiology has no limits. It deals with similarities and dissimilarities of organ functions and integration of organ systems. Thus, I wish to compare time-dependent, thermoregulatory responses of a dog to thermal stress with time-dependent responses of a Pekin duck to salt stress. In this comparison, we can recognize similarities, in that both illustrate the phenomenon known as *homeostasis embracing negative feedback*. We shall also recognize an important dissimilarity. I shall argue that the response of the nasal salt glands of the Pekin duck to a hypertonic intravenous (i.v.) infusion of sodium chloride illustrates a new phenomenon, homeostasis embracing negative feedback *enhanced and sustained by positive feedback*.

Thermal Regulatory Responses to Thermal Stress

Steady-state responses to thermal stress have been investigated in many species of endotherms as well as ectotherms. Measurements usually include body temperatures and oxygen consumption, an indirect measure of the rate of heat production. However, measurements of body temperatures and rates of heat production and heat loss are rare during the transition interval between initial response and steady-state response to thermal stress. Fig. 15.1 will illustrate these responses in an 8.5-kg dog. A trained and resting dog was moved from a neutral environment and placed in a gradient layer calorimeter for 5 hr. On separate days, the calorimeter was maintained at 9°, 25°, and 35°C. A ventilated hood was placed over the dog's head for measurements of rates of oxygen consumption and evaporation of water from the head. Rates of heat production and respiratory heat loss were calculated from these measurements. The rate of dry heat loss was calculated from the voltage output of the gradient layer.

In the cold calorimeter, the initial rate of heat production was the same as in a neutral environment, whereas the rate of heat loss was more than

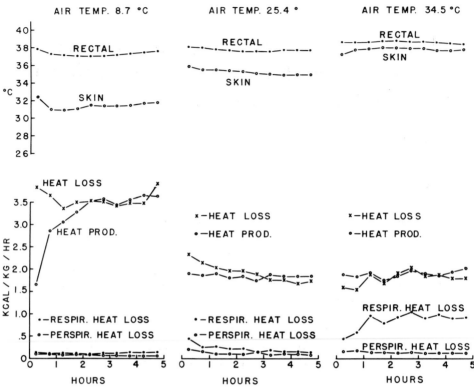

FIGURE 15.1. Rates of heat production and heat loss in an 8.5-kg dog in response to a transition from a neutral environment to a cold environment (*left*), a neutral environment (*middle*), and a hot environment (*right*). (From Hammel et al. 1958.)

double that rate. This difference indicates that heat stored in the body in the neutral environment was being lost to the cold calorimeter. Consequently, skin and core temperatures were falling during the initial period in the calorimeter. Shivering and the rate of heat production increased. This response may be attributed to the lower core and skin temperatures. Body temperatures continued to fall, and the rate of loss of stored heat continued to be greater than zero for about 2 hr and until the rate of heat production increased to become equal to the rate of heat loss. For the remaining 3 hr, the rates of heat production and heat loss were about equal. Actually, it is puzzling that heat production slightly exceeded heat loss so that skin and core temperatures slowly increased. We were not concerned with this discrepancy when these data were obtained. These three experiments confirmed our expectation that thermal homeostasis (homeothermy) was attained in the dog by activation of appropriate responses to thermal stress, and to an appropriate extent such that the thermal stimuli actuating the responses were no longer changed by the thermal stress.

Proportional Control with Adjustable Reference

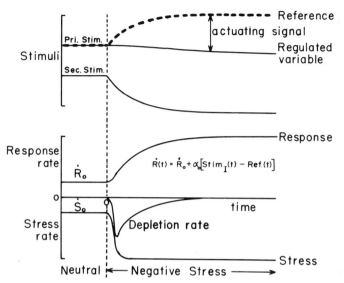

FIGURE 15.2. Temporal relationships between stress rate, stimuli, and response rate are depicted. These illustrate homeostasis embracing negative feedback alone. An essential feature of this model is the quantity lost during the time the stress rate exceeds the response rate. The depletion rate is the stress rate minus the response rate, and its time integral equals the deficit. Reduction in the associated intensive properties are transduced by receptors to stimuli which actuate the response. The model incorporates the concept that a stimulus derived from a receptor which responds rapidly to the stress participates in the actuating signal by offsetting the reference. Thereby, the regulated variable deviates less during the stress (Hammel et al. 1963; Yates et al. 1961.)

A regulatory response to a negative stress may be represented and interpreted as illustrated in Fig. 15.2. A generalized extensive quantity resides in the body, and the amount of it can be characterized by an intensive property of the body times the mass or volume of the body. The rate of production of this quantity in the body in a neutral environment is $\dot{R}_0$, and the rate of its loss to the neutral environment is $\dot{S}_0$. In the neutral environment, $\dot{R}_0 = -\dot{S}_0$, and both remain constant in time so that the corresponding intensive property remains constant. When a constant negative stress is imposed at time 0, $-\dot{S}(t)$ greatly exceeds $-\dot{S}_0$ and $\dot{R}_0$ as well, so that some of the quantity stored in the body is lost, and the associated intensive property of parts of the body decreases. These changes may be transduced by sensory receptors and interoceptors into neural signals which are stimuli to actuate an appropriate response, $\dot{R}(t)$. The response is appropriate in kind when its effect on the intensive property is opposite in sign to the effect of the stress. Feedback is negative for an appropriate response. Feedback is positive for an inappropriate response so that the intensive property continues to deviate, as caused by both the stress and the response. Homeostasis is attained when the response is appropriate, and the response rate becomes equal to the rate of stress. When $\dot{R}(t) = -\dot{S}(t)$, the associated intensive properties of all parts of the body attain steady state. Homeostasis embracing negative feedback alone requires a deficit in the body of the quantity lost during negative stress. This body deficit ensures adequate stimuli to sustain the response. Furthermore, the rate of response can exceed the rate of stress only transiently and never for a time sufficient to eliminate the body deficit.

Osmoregulatory Response to a Salt Stress

The kidneys of marine birds are unable to clear their blood of excess sodium chloride when the birds ingest marine invertebrates. A pair of orbital salt glands perform this function in these birds (Schmidt-Nielsen 1960). The secretory nerve is preganglionic and goes to the ethmoid ganglion attached to the gland. Postganglionic fibers pass to the peritubular cells of secretory tubules (Fänge et al. 1958). In response to acetylcholine, blood flow to these cells increases, and they transport NaCl from adjacent interstitial fluid to the lumen of the tubules. Luminal fluid is collected and flows to a vestibule near the external nares where it is discharged (Peaker and Linzell 1975). As in the preceding section, we are interested in the process whereby homeostasis is attained—how an excess of salt in the body is recognized so that a signal can actuate its elimination. If this homeostatic process were to embrace negative feedback alone, then the rate of salt excretion in relation to the rate of salt loading would be as depicted in Fig. 15.3. Since Na^+ and Cl^- are primarily extracellular ions, the amount of salt in the body is roughly the product of the salt concentration and volume of the extracellular fluid (ECF). It has been shown (Deutsch et al. 1979; Hammel et al. 1980; Kaul and Hammel 1979), and will be confirmed in the next figure, that increasing the tonicity and/or the volume of the ECF elicits excretion by the salt glands, as depicted in Fig. 15.3. In this representation of the response of the salt glands to an intravenous infusion of hypertonic salt stress, salt accumulates in the body, increasing both the tonicity and volume of the ECF. Interoceptors transduce some intensive properties of these changes into stimuli, and these stimuli cause the actuating signal which is transmitted via the secretory nerves to the salt glands.

The actual response of an average Pekin duck to an i.v. infusion of 1000 møsm NaCl $kg^{-1}H_2O$ at 0.4 ml min^{-1} for 60 min is shown in Fig. 15.4, infusion A. The duck had been excreting a prior infusion but was no longer excreting. It was at threshold when infusion A commenced. During the first 15-min period, the infusion rate exceeded the excretion rate, so salt was accumulating in the ECF. Since the concentration of the infusate

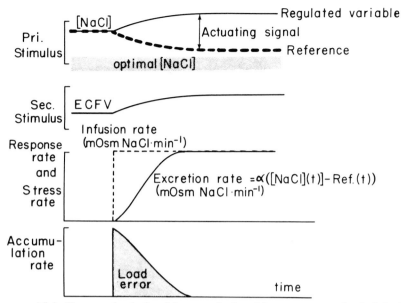

FIGURE 15.3. Temporal relationships are depicted between rate of salt infusion, changes in salt concentration and volume of extracellular fluid, and rate of excretion. Salt accumulates in the ECF, increasing both tonicity and volume of the ECF, and these changes stimulate the response of the salt glands.

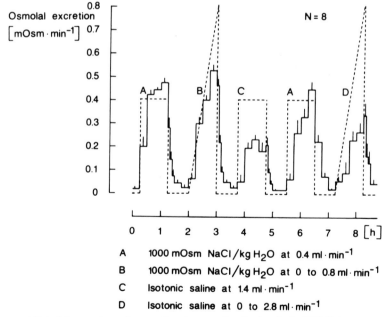

A	1000 mOsm NaCl/kg H_2O at 0.4 ml·min^{-1}
B	1000 mOsm NaCl/kg H_2O at 0 to 0.8 ml·min^{-1}
C	Isotonic saline at 1.4 ml·min^{-1}
D	Isotonic saline at 0 to 2.8 ml·min^{-1}

FIGURE 15.4. Mean rates of excretion by nasal salt glands ($\pm$SEM) in response to 1-hr i.v. infusions of hypertonic and isotonic saline solutions infused as square and ramp infusions. After each infusion, excretion returned to near zero so that stimuli were at threshold levels before the next infusion.

was chosen to match the concentration of the gland excretion, about 1000 møsm kg^{-1}H$_2$O, both the tonicity and the volume of the ECF increased. The ECFV increased for two reasons: (1) fluid was added to the plasma at 0.4 ml min^{-1}, and (2) since the osmolality of the ECF was increased, water was rapidly drawn from intracellular fluid such that the osmolality of all body water increased at nearly the same rate.

During the next 15-min period of infusion A, the excretion rate already exceeded the infusion rate. This excessive excretion rate continued to the end of the infusion. Before the infusion ended, while the rate of excretion remained excessive, at least as much salt had been excreted as was infused. The amount of salt in the ECF, as the excretion remained excessive, was the same as or even less than it had been in the beginning before excretion began. This response is shown again in Fig. 15.5 for an individual duck. The initial osmolality of a blood sample taken before infusion A, was used to calculate subsequent changes in ECF tonicity and

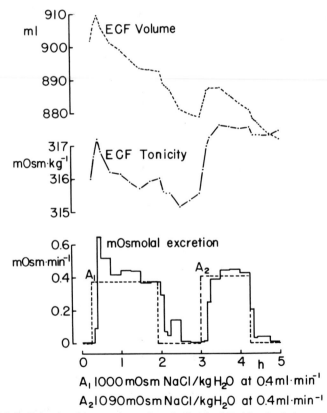

A_1 1000 mOsm NaCl/kg H$_2$O at 0.4 ml·min^{-1}

A_2 1090 mOsm NaCl/kgH$_2$O at 0.4 ml·min^{-1}

FIGURE 15.5. Salt gland excretion of an individual Pekin duck in response to a 100-min infusion of hypertonic saline solution. ECF volume and tonicity are calculated based on gain and loss of salt and water throughout the experiment.

volume. Account was taken of all salt and water in and out of the duck at the end of each 15-min interval. It was assumed that water was rapidly exchanged between the intracellular and extracellular fluid compartments for all tissue. Again, in less than an hour, the tonicity and volume of the ECF were less than at threshold; yet, the excretion rate continued for another hour to exceed or equal the infusion rate. Should infusion continue for many hours, the excretion rate would continue to equal or even exceed the infusion rate.

Infusion C, Fig. 15.4, illustrates that increasing the volume of the ECF alone also stimulates excretion. Infusion C was 0.9% NaCl; its osmolality was 287 møsm NaCl $kg^{-1}H_2O$. The average osmolality of the plasma at threshold was 324 ± 5 møsm $kg^{-1}H_2O$. Thus, infusate C was slightly hypotonic to the duck's plasma and could not have increased the tonicity of the ECF. Infusion C, during the first 15 min, increased the volume of the ECF by about the same percentage (3%) as infusion A during the first 15 min. Since there was no concomitant increase in ECF tonicity, the excretion rate elicited by infusion C was somewhat less than half the excretion rate elicited by infusion A. Since the duck's excretion was about 1000 møsm $kg^{-1}H_2O$ for the next 45 min, it excreted less water than it received. Thus, the ECF tonicity was steadily decreasing due to this hydration. The only stimulus to which the excretion can be attributed during infusion C is the increasing volume of the ECF. We may conclude that about half the initial stimulus derived from infusion A during the first 15 min was from increasing ECF tonicity and the other half was from increasing ECF volume. However, the sustained excretions in response to infusion A and C cannot be attributed to these changes alone for the remaining 45 min of the 1-hr infusion.

Interpreting a Response Which Exceeds the Stress

It is not usual for the response rate to exceed the stress rate. And rarely does the integrated response exceed the integrated stress because that eliminates the stimuli which initiate the response. If the stimuli return to or fall below their threshold values, what process sustains the response? First, we might ask whether we have misinterpreted the effects of the response on the stimuli. We have assumed that as the infusate enters the bloodstream, its effect on the blood is distributed throughout the extracellular fluid within a few circulation times (certainly less than 3 min). We have further assumed that a change in the tonicity of the ECF affects the flow of water into or out of the cells of the body's tissue with similar rapidity. That is, the intracellular and extracellular fluids are equilibrated in less than a few minutes.

Rapid equilibration does not imply that the fluid spaces of all tissues have the same compliance or that they are the same percentage of the tissue water, rather only that their fluid spaces exchange water with the same rapidity or within a few minutes. The ECF volume of salt-acclimated Pekin ducks is about 26% of the body weight (body volume) as indicated by the inulin space (Ruch and Hughes 1972) or by the sodium space (Kaul and Simon 1983). This percentage is only an average for all tissues. It is likely that the volume of the interstitial space as a percentage of tissue volume varies from organ to organ, and it is also likely that the compliance of the interstitial space varies from organ to organ. This latter consideration may be important because it has been shown that it is the change in interstitial fluid volume that affects salt gland excretion, not a change in plasma volume. Adding a colloid (Dextran 70) to the hypertonic NaCl solution infusing a duck lowers the hematocrit and inhibits salt gland excretion (Hammel et al. 1980). We may speculate that a stretch receptor transducing changes in the interstitial volume would most likely be situated in tissue with a high compliance within its interstitium. Thus, increasing the tonicity of the ECF and moving water out of the intracellular fluid into the ECF would affect a larger change in the volume of an interstitium with a high compliance. We also speculate that a tonicity receptor is an interoceptor which responds to loss of water from its interior. A change in the polarization of its membrane may be due to a change in receptor volume or to a change in the intracellular-to-extracellular ratio of some ion. The location of these tonicity and interstitial volume receptors is not known. It is known that afferents in the vagii are important if not essential for salt gland function (Hanwell et al. 1972; Gilmore et al.). It is also known that changing the concentration of NaCl in carotid blood has a larger effect on salt gland excretion than a similar change in venous blood (Hammel et al. 1983). This suggests that receptors sensitive to NaCl reside in the brain.

RESPONSES TO INCREASING AND DECREASING ECF TONICITY AND VOLUME ARE RAPID

Rapid response to changes in plasma tonicity has been demonstrated in the head and very likely the brain (Hammel et al. 1983). Injecting a 2-ml bolus of 1000 møsm NaCl kg^{-1} into carotid blood has a larger effect on salt gland excretion than the same bolus injected intravenously. The intracarotid (i.c.) effect is larger than the i.v. effect whether the duck is receiving a continuous i.v. infusion of 1000 møsm NaCl $kg^{-1}H_2O$ at 0.4 or at 0.2 ml min^{-1}, or whether the duck is at threshold after no infusion or at subthreshold receiving an i.v. infusion of 200 møsm glucose $kg^{-1}H_2O$ at 1.0 ml min^{-1}. Moreover, the i.c. response is larger than the i.v. response during the first 5 min in all cases and also during the second 5 min, when the duck is at threshold or subthreshold. An intracarotid bolus infusion of a 2-ml solution of 200 møsm glucose $kg^{-1}H_2O$ into a duck excreting an i.v.

infusion of 1000 mOsm NaCl kg^{-1} at 0.4 ml min^{-1} inhibits excretion more than the same 2-ml bolus of water-infused i.v. An i.c. infusion of the glucose solution at 1 ml min^{-1} for 15 min reduces, and may stop, excretion within 3 min, and excretion remains zero for the remaining 10 or more minutes. Thus, whether the change in tonicity of the blood causes water to move out of or into the sodium chloride receptors in the brain, the response of the salt glands is within a few minutes. We can conclude from this that an excretion of salt in excess of the salt infused described in the previous section may not be ascribed to sluggish movement of water into or out of the cells of the tonicity receptors which reside within and/or outside of the central nervous system.

INTEGRAL CONTROL UNLIKELY

Integral control comes to mind as a controlling system which has the feature to eliminate the load error. Suppose that the central nervous system had the ability to integrate, in time, deviations of the stimuli from their threshold values and then generate an actuating signal in proportion to these integrals. Formally, the response rate $\dot{R}(t)$ could be expressed

$$\dot{R}(t) = \alpha \int_{t_0}^{t} [T(t - T_0)]dt + \beta \int_{t_0}^{t} [V(t) - V_0]dT \qquad (15.1)$$

where $T(t)$ is the ECF tonicity at time t, $V(t)$ is the ECF volume at time t, T_0 and V_0 are their respective threshold values prior to time $t = t_0$, α and β are proportionality constants, and $\dot{R}(t_0) = 0$ when $T(0) = T_0$ and $V(0) = V_0$. Considering infusion A in Fig. 15.4, $T(t)$ and $V(t)$ increase during the first 15 min because the infusion rate exceeds the excretion rate, and then they start to decline during the next 15 min because the excretion rate now exceeds the infusion rate; yet all the time, $T(t)$ and $V(t)$ exceed T_0 and V_0, respectively. The sum of these integrals would produce a response at 30 min, $\dot{R}(30)_A$. Now consider infusion B in Fig. 15.4; again, $T(t)$ and $V(t)$ increase during the first 15 min because the infusion rate exceeds the excretion rate, and they start to decline during the next 15 min.

In both infusions A and B, if the response rate depends on the magnitude of the integrals, then these depend on how much the average response rate is less than the average infusion rate. In infusion A, the average response rate was 78% of the average infusion rate for the first 30 min, whereas in infusion B the average response rate was 91% of the average infusion rate. Assuming the same integral control for infusions A and B, the lesser response in A should have generated a larger stimulus to excrete. Since it did not and since the responses were not the same in both infusions, this suggests that time integration of the actuating signal is not the way the central nervous system functions to cause excretion in excess of infusion.

A Positive Feedback Loop Is Feasible and Plausible

Maintenance of an intensive property of the body within a narrow range of variation requires responses which limit the increase of the property as well as responses which limit its decrease. Each response must be inhibited until the intensive property deviates from a threshold value for that response. Thus, neurons which stimulate salt gland excretion must be inhibited until threshold values of ECF tonicity and volume are exceeded. These threshold values are already in excess of threshold values eliciting release of the antidiuretic hormone arginine vasotocin, thirst, and water drinking. On the other side, there are threshold values of ECF tonicity and ECF volume below which salt retention by the kidneys and salt appetite are elicited. This leads to the concept that receptors and interneurons which facilitate the class of appropriate responses must send collateral axons to inhibit the opposite class of responses which would be inappropriate. Collateral axons are an integral part of regulatory neural networks, and in these cases they are inhibitory.

It is feasible and plausible that, in some situations, the collateral axons facilitate. Suppose that a neuron in the pathway leading to the secretory nerve of the nasal salt gland has a collateral axon that releases a neuromodulator or transmitter substance which facilitates the same neuron or similar neurons in the same nuclei. Such a feedback loop would be positive in that, once stimulated, the neuron would continue to stimulate itself. It could continue its activity even as the stimuli which initiated its activity vanish. Such a neuron would enhance and sustain a response, once initiated, until it was inhibited. The neuron is already endowed with inhibitory input so that it becomes inhibited as ECF tonicity and volume fall below threshold values. As it is inhibited, its autofacilitation is diminished. Such a positive feedback loop is rapidly shut off when infusion of the salt load ceases, so that excretion is also rapidly terminated.

Summary

Induction of thermoregulatory responses to thermal stress reveals no compelling evidence that negative feedback alone was insufficient to describe the results. Homeothermy was sustained by sustained deviations of some or all body temperatures from their respective values in the neutral environment. In contrast, the salt excretion rate exceeded the infusion rate to an extent and for a time sufficient to eliminate the initial stimuli. Nevertheless, the excretion rate continued to exceed or equal the infusion rate until infusion ceased and excretion was rapidly terminated. In this instance, homeostasis embraced negative feedback to ensure activation of the appropriate response. In addition, a positive feedback loop may be appended to the neural network which could enhance and sustain the response after the initial stimuli have vanished. We may consider whether this process may also enhance and sustain other responses in regulatory biology when their stimuli appear insufficient to sustain the response. We might anticipate that a positive feedback loop would enhance responses to severe stress.

Acknowledgments. Research on salt gland function was conducted in part at the Max-Planck-Institut für physiologische und klinische Forschung, W.G. Kerckhoff-Institut, 0-6350 Bad Nauheim, Fed. Rep. Germany. It was also supported in part by National Science Foundation Grant PCM 82-12072. The content of this chapter was presented as the Fifth Annual Irving-Scholander Memorial Lecture on September 25, 1985, at the University of Alaska, Fairbanks. Walter Klein designed and constructed the ramp infusor used in infusions B and D, Fig. 15.4.

REFERENCES

Deutsch H, Hammel HT, Simon E, Simon-Oppermann CH (1979). Osmolality and volume factors in salt gland control of Pekin ducks after adaptation to chronic salt loading. *J Comp Physiol* 129: 301–308

Fänge R, Schmidt-Nielsen K, Robinson M (1958). Control of secretion from the avian salt gland. *Am J Physiol* 195: 321–326

Gilmore JP, Gilmore C, Dietz J, Zucker IH (1977) Influence of chronic cervical vagotomy on salt gland secretion in the goose. *Comp Biochem Physiol* 57A: 119–121

Hammel HT, Jackson DC, Stolwijk JAJ, Hardy JD, Stromme SD (1963). Temperature regulation by hypothalamic proportional control with an adjustable set point. *J Appl Physiol* 18: 1146–1154

Hammel HT, Simon-Oppermann CH, Simon E (1980). Properties of body fluids influencing salt gland secretion in Pekin ducks. *Am J Physiol* 239: R489–R496

Hammel HT, Simon-Oppermann CH, Simon E (1983). Tonicity of carotid blood influences salt gland secretion in Pekin ducks. *J. Comp Physiol* 149: 451–456

Hammel HT, Wyndham CH, Hardy JD (1958). Heat production and heat loss in the dog at 8–36°C environmental temperature. *Am J Physiol* 194: 99–108

Hanwell A, Linzell JL, Peaker M (1972). Nature and location of the receptors for secretion by the salt gland of the goose. *J Physiol (Lond.)* 226: 453–472

Kaul R, Hammel HT (1979). Dehydration elevates osmotic threshold for salt gland secretion in the duck. *Am J Physiol* 237: R355–R359

Kaul R, Simon E (1983) Apparent volume of distribution of Na in Pekin ducks. (abstr.) *Naunyn-Schmiedeberg's Arch* 322: R19

Peaker M, Linzell JL (1975). *Salt Glands in Birds and Reptiles.* Cambridge University Press, Cambridge, London

Ruch, FE, Hughes MR (1972). The effects of hypertonic sodium chloride injection on body water distribution in ducks (*Anas platyrhyncos*), gulls (*Laurus glaucescens*) and roosters (*Gallus domesticus*). *Comp Biochem Physiol* A52: 21–28

Schmidt-Nielsen K (1960). Salt secreting gland of marine birds. *Circulation* 21: 955–967

Yates FE, Leeman SE, Glenister DW, Dallman MF (1961). Interaction between plasma corticosterone concentration and adrenocorticotropin-releasing stimuli in the rat: Evidence for the reset of endocrine feedback control. *Endocrin* 69: 67–80

16

Time, Energy, and Body Size

Hermann Rahn

Different mammals and birds appear to be remarkably similar in their function but differ in their overall geometry as they increase in size. As Brody (1945) stated so well, "The organism changes geometrically so as to remain the same physiologically." In this context I want to examine these changes in terms of the allocation of time and energy as animals increase in size. Such considerations suggest that during the life span of mammals the total resting energy expenditure per unit body mass is similar, a concept that can be traced back to Max Rubner at the turn of the century. It also suggests, for example, that the number of heart beats and breaths during a life span are similar from mouse to elephant.

The evidence for such provocative statements is not new and is based on interpretations of allometric relationships which examine organ size, metabolic rates, physiological rates, and cycles as they vary with body size, and have most recently been discussed by Boddington (1978), Calder (1984), Lindstedt (1985), Lindstedt and Calder (1976, 1981), and Schmidt-Nielsen (1984).

As animals become larger, their physiological functions, or biological machinery, are remarkably similar, preserving throughout evolution a more or less constant cell size, capillary diameter, capillary distance from cells, blood pressure, electrolyte composition, and Starling's balance between hydrodynamic and osmotic forces. Similar are muscle filaments as well as their maximal force per cross-sectional area, body temperature, alveolar gas tensions, and blood pH, along with efficiencies of energy utilization.

While these dimensions and compositions are preserved, blood volume, heart volume, lung volume, and muscle volume must increase with size and do so in direct proportion to body mass.

However, energy production and the volume flows which provide for O_2 delivery and CO_2 elimination, the alveolar ventilation, and cardiac output increase as the ¾ power of body mass. That is, for every 10-fold increase in body mass, these functions increase only 5.6-fold.

Lastly, we must look at what I call delivery systems. These decline as animals get larger, namely, heart rate, breathing rate, cardiac output per kilogram, alveolar ventilation per kilogram, specific metabolic rate (kilocalories per kilogram), and some renal functions. As the recent review of Lindstedt and Calder (1981) reveals, these functions all appear to decline at similar rates; namely, they are proportional to ca. -0.25 power of body mass. Some of these functions are more conveniently expressed in terms of cycles or time, such as cardiac cycle, respiratory cycle, life span, circulation time, and gestation time. These are all proportional to ca. $+0.25$ power of body mass. The mass exponents for some of these functions are shown in Table 16.1.

Most of these specific generalizations were unknown to Max Rubner, professor of physiology at Berlin and director of the Kaiser Wilhelm Institute for Work Physiology and Nutrition around the turn of this century (Fig. 16.1). Rubner, a powerful scientist of this era, had finally established the fact that the heat production of animals measured by direct calorimetry could be explained by the combustion of food and also measured indirectly by the uptake of O_2. Today this is an accepted fact. In 1908 and 1909 toward the end of his career he published two books on the relationship between life span, growth, and nutrition. In them he presented a table which showed the body mass and estimated life span of five domestic animals. The last column showed the total calories per kilogram expended during the life span (basal metabolic rate per kilogram × life span), and pointed out that these were essentially similar (Table 16.2). He was quick to note that humans did not fit into this scheme. He estimated the life span of the human to be 80 years, and his estimated total caloric expenditures per kilo was four times greater than that for all the other animals. (I shall return to this point later.) In any event, at that time to suggest that the lifetime caloric expenditure per unit body mass was similar in animals small and large was a very bold statement, and there it rested. During the following decades most biologists were skeptical since the data base was rather poor. Furthermore, estimates of life span are a rather elusive figure. Nevertheless, the concept was attractive, and over the following decades the data base increased.

TABLE 16.1. Specific rates or cycle times and their exponents when regressed upon body mass.

Rates		Time	
Specific metabolism	$- 0.25$	Life span	0.24
Heart contraction	$- 0.23$	Cardiac cycle	0.23
Breathing frequency	$- 0.26$	Respiratory cycle	0.26
Cardic output (kg^{-1})	$- 0.20$	Gestation period	0.25
Alveolar ventilation (kg^{-1})	$- 0.26$	Circulation time	0.21

Selected from the tables of Lindstedt and Calder (1981).

FIGURE 16.1. Max Rubner, appointed professor of physiology at Berlin, 1908, and founder of the Kaiser-Wilhelm Institute for Work Physiology. (Photo from Geschichte der Physiologie, K.E. Rothschuh, Springer, 1953.)

TABLE 16.2. Relationship between body mass, life span, and metabolic output presented by Rubner in 1908.

	Body weight	Length of life	kcal
	(kg)	(years)	kg·life span
Horse	450	30	170,000
Cow	450	26	141,000
Dog	22	9	164,000
Cat	3	8	224,000
Guinea pig	0.6	6	266,000

Metabolic Rate. One of the most reliable and accepted allometric relationships for mammals is that the basal metabolic rate is proportional to the body mass raised to the 0.75 power (Kleiber 1961). Dividing the metabolic rate by body mass we obtain the specific metabolic rate which now declines with an increase in body mass raised to the −0.25 power, which tells us, for example, that a 3-g shrew burns its fuel about 20 times faster than a 4000-kg elephant.

Life Span. Our exact knowledge of life spans for various mammals is not easily established, and I have combined the allometric equations for life span as a function of body mass from the reports of Boddington (1978), Gunther and Guerra (1955), and Sacher (1959), which show that life span is proportional to the +0.25 power of body mass, in contrast to the −0.25 power which describes specific metabolic rate. In Fig. 16.2 these two functions are shown on a semilog plot where the specific metabolic rate is expressed per year and life span is expressed in years. These two reciprocal curves tell us that for any given body mass the product of specific metabolic rate and life span yields a constant, namely, 240,000 kcal·kg^{-1} (240 kcal·g^{-1}), a value not too different from that predicted by Max Rubner 80 years earlier (Table 16.2). A recent study by Boddington (1978) yields a similar prediction and refers to this constant as the *absolute metabolic scope* of mammals.

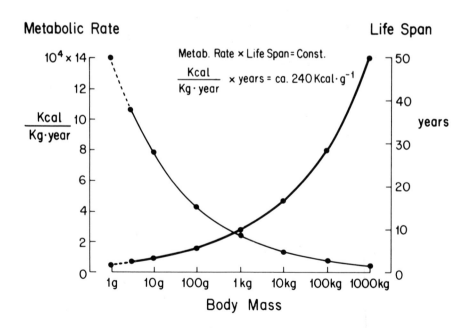

FIGURE 16.2. Relationship of specific metabolic rate and life span to body mass in mammals.

It is of interest to note that the metabolic rate of a 70-kg man fits well on the metabolic rate curve. However, the life span curve predicts a life span of ca.25 years, which was the life expectancy of man prior to the seventeenth century (Brody 1945). The fact that humans have been able to rid themselves over the last millennium of many factors which limit the life span of other animals, such as predation, disease, and deleterious nutritional and environmental conditions, might explain man's extended period.

Heart and Respiratory Rate. The heart rate of mammals has been well established and is also proportional to the -0.25 power of body mass. In Fig. 16.3, heart rate and life span are plotted against body mass, which tells us that during the life span of most mammals the heart beats about 1.2 billion times. A similar relationship can be established for the respiratory frequency, which yields an average predicted number of 300,000 per life span. Thus, on the average, from mouse to elephant, there are four heart beats for each respiration.

The Typical 1-g Tissue

The typical 1-g tissue illustrated in Fig. 16.4. I call it the 1-g mammalian tissue "on the hoof." It represents all mammals, mouse to elephant. During its life span it burns the equivalent of 60 g of carbohydrates to

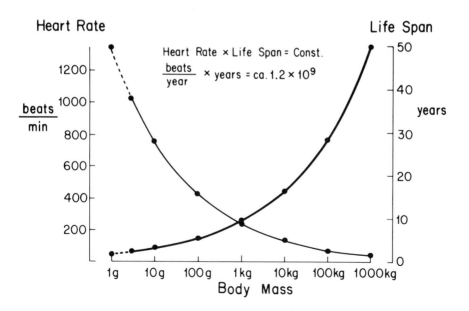

FIGURE 16.3. Relationship of heart rate and life span to body mass in mammals.

release 240 kcal. This requires 50 l of O_2, and with an RQ of 0.8 will release 40 l of CO_2.

We can describe additional functions. For example, what is the perfusion that this 1 g of tissue receives during its life span? Cardiac output per kilogram of body mass has a similar negative slope (Table 16.1). So during a life span mammals receive about 800 l of blood per gram of tissue. Since this tissue also consumes 50 l of O_2, it is easy to calculate the arterial-venous O_2 difference, namely, 6 vol %. We also recall that it took about 1 billion heart contractions to pump this blood.

Before I launch into a critique of this recapitulation, let me bring you, for your possible amusement, another example of life span with which you may feel more at home. I began to wonder whether our modern engineer had succeeded in developing an organism which was more efficient than our typical mammalian tissue. I am referring to the American car. I assumed it weighed 2000 kg, had an average life span of 40,000 miles, and obtained 20 miles per gallon of gasoline. In Fig. 16.4 you

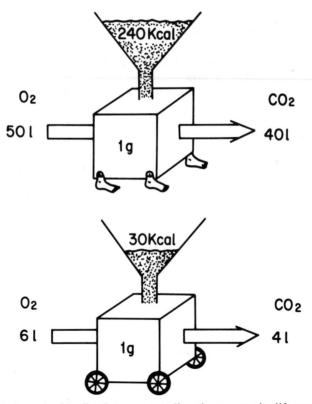

FIGURE 16.4. *Top:* An idealized 1-g mammalian tissue over its life span metabolizing 240 kcal of food by taking up 50 l of O_2 and producing 40 l of CO_2. *Bottom:* Metabolic performance of an idealized 1 g of American automobile over its life span of 40,000 miles.

see 1 g of American car "on wheels." Note that it consumed only one-eighth the amount of fuel before it landed in the junkyard. In other words, without the continuous repair that our tissues undergo, the engineer cannot compete, as yet, with the efficiency of protoplasm.

Critique

There are, of course, many questions which arise from this oversimplified presentation of "Rubner's law." Nevertheless, it does indicate a general trend even though there are many species that do not fit on the idealized curves that were presented. For example, (1) Scholander et al. (1950) showed that certain species had metabolic rates that deviated significantly from the common 0.75 power function of body mass. Would these deviations be associated with a different life span? (2) Rubner's law is based on basal metabolic rates, while these, of course, are normally exceeded during the daily activity of animals. Should, therefore, this overall metabolism be included in such calculations? And (3), finally, there is the question concerning the reliability of life-span estimates. These questions cannot be easily answered; instead I would like to provide other evidence in support of Rubner's law.

Metabolic Rates and Life Span in Birds

Metabolic rates and life span in birds were reviewed by Lindstedt (1985) and Lindstedt and Calder (1976). From these, one can calculate that the total energy per gram of tissue during the life span of passerine birds is also a constant, but considerably higher than in the mammal, namely, ca.1000 kcal·g^{-1} instead of 240 kcal·g^{-1}. The explanation for this difference is not obvious, but the metabolic rate as well as the life span is higher than for a mammal of the same body mass. Nevertheless, it provides additional support for Rubner's law, and in this case the life span data were determined from extensive bird-banding records, which are more reliable than those established for mammals.

Embryonic Life Span

Can the overall life span be conveniently divided into three periods, namely, embryonic, juvenile, and adult, where each period functions energetically as described above for the adult mammal? This was originally proposed by Rubner (1909) and tested more recently by Rahn (1982) and Rahn and Ar (1980). The advantage that developing bird eggs provide in this context is that it is relatively easy to measure their oxygen consumption throughout the incubation period and that the incubation times are well established.

One can now ask the question: Is the total energy expenditure per gram of egg during incubation the same in all bird eggs, even though there are large differences in egg mass and incubation time? Do bird embryos have an *absolute metabolic scope* in the sense of Boddington (1978)? The total incubation energy expenditure has recently been measured in a large number of species (see Bucher and Bartholomew 1984 for major source). These values are regressed against egg mass in Fig. 16.5 and have a slope of 0.95, not significantly different from 1.0. The average energy loss was 2.3 kJ/g egg (0.55 kcal/g egg), where egg mass ranged from 1 g to the 1450-g ostrich egg (not shown in the figure), and incubation times ranged from 11 days to 65 days in two species of albatross. Since the hatchling mass is 67% of the initial egg mass, we can convert the total energy requirements of the embryo to hatchling mass. Thus, 0.82 kcal/g hatchling is the best estimate of the *absolute metabolic scope* during the embryonic life span. Furthermore, the gas tensions which exist in the air cell just

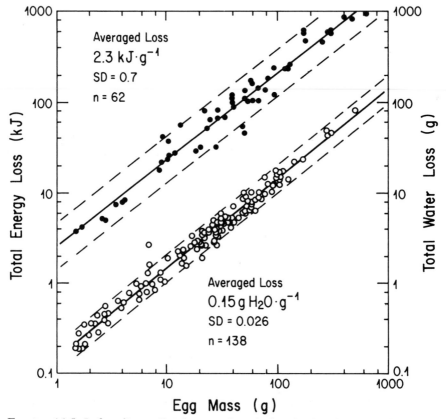

FIGURE 16.5. *Left ordinate:* Total energy loss during the development of an avian embryo as a function of the initial egg mass. *Right ordinate:* Total diffusive water loss during the development of avian eggs as a function of initial egg mass.

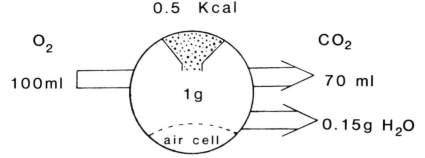

FIGURE 16.6. Idealized 1-g bird egg over its embryonic life span, metabolizing 0.55 kcal of stored energy by taking up 100 ml O_2, producing 72 ml CO_2, and releasing 0.15 g of water vapor.

before the first pipping stage are similar, namely, $P_{O_2} = 100$ Torr and $P_{CO_2} = 42$ Torr (Paganelli and Rahn 1984). These have now been directly measured in some 22 species ranging in size from 10 to 1500 g (ostrich) and with incubation periods varying from 15 to 65 days.

Also shown in Fig. 5 is the total diffusive water loss of eggs (daily water loss times total incubation days) during their natural incubation (Ar and Rahn 1980). The slope of this line is 0.99 and not significantly different from 1.0. Thus the average loss is 0.15 g $H_2O g^{-1}$ egg. This water is lost by diffusion through the pores of the eggshell and does not include the additional amount that is lost by convection after pipping of the eggshell occurs.

Fig. 16.6 depicts the idealized 1-g avian egg during its embryonic life span. It delivers a hatchling with similar caloric loss, water loss, and gas tensions, which are independent of incubation time or mass.

Summary

In 1908 Rubner suggested that during the life span of mammals their total basal energy expenditure per kilo body mass is similar, namely, ca.200,000 kcal, a concept which is here reexamined. Today's evidence still fits this model and can be extended to adult passerine birds, while energy expenditure per unit mass of developing avian eggs also fits Rubner's law. The latter suggests that total life span can conveniently be divided into embryonic, juvenile, and adult life spans where during each period the total specific energy expenditure is similar.

As J.B.S. Haldane (1985) wrote, "For every kind of animal there is a most convenient size," to which we can add "and an optimal metabolic rate." Each gram of tissue, from mouse to elephant, is provided with the same potential for energy release, which can be spent quickly or slowly depending on the animal's size. The larger the animal, the smaller its maintenance cost and the longer the life span.

It may not be surprising that today Rubner's "law" has become the focus for one of our theories of aging, the *free radical* theory. It has been estimated that during tissue metabolism as much as 6 percent of the oxygen is converted to various reactive O_2 species, which in turn are countered by large numbers of antioxidants. Thus one might argue that after a life span delivery of 50 l of O_2 per gram of average tissue, the balance between reactive O_2 species and antioxidants is upset leading to the stage of "burn-out."

REFERENCES

Ar A, Rahn H (1980). Water in the avian egg: Overall budget of incubation. *Am Zool* 20: 373–384

Boddington MJ (1978). An absolute metabolic scope of activity. *J Theor Biol* 75: 443–449

Brody S (1945). *Bioenergetics and Growth* (rpt. 1968). Hafner, New York

Bucher TL, Bartholomew GA (1984). Analysis of variation in gas exchange, growth pattern, and energy utilization in a Parrot and other avian embryos. In: Seymour RS (ed) *Respiration and Metabolism of Embryonic Vertebrates*. Dr. W. Junk Publishers, Dordrecht, pp 359–372

Calder WA (1984). *Size, Function and Life History*. Harvard University Press, Cambridge, MA

Gunther B, Guerra E (1955). Biological similarities. *Acta Physiol Latin-Amer* 5: 169–186

Haldane JBS (1985). In: Smith JM (ed) *On Being the Right Size and Other Essays*. Oxford University Press, Oxford

Kleiber M (1961). *The Fire of Life*. John Wiley & Sons, New York

Lindstedt SL (1985). Physiological time and aging in birds. In: Lints FA (ed) *Non-mammalian Models for Research in Aging*. Karger, Basel

Lindstedt SL, Calder WA (1976). Body size and longevity in birds. *Condor* 78: 91–94

Lindstedt SL, Calder WA (1981). Body size, physiological time, and longevity of homeothermic animals. *Quart Rev Biol* 56: 1–16

Paganelli CV, Rahn H (1984). Adult and embryonic metabolism in birds and the role of shell conductance. In: Seymour RS (ed) *Respiration and Metabolism of Embryonic Vertebrates*. Dr. W. Junk Publisher, Dordrecht, pp 192–204

Rahn H (1982). Comparison of embryonic development in birds and mammals: Birth weight, time, and cost. In: Taylor CR, Johansen K, Bolis L (eds) *A Companion to Animal Physiology*. Cambridge University Press, Cambridge, pp 124–137

Rahn H, Ar A (1980). Gas exchange of the avian egg: Time, structure and function. *Am Zool* 20: 477–484

Rubner M (1908). *Das Problem der Lebensdauer und seine Beziehungen zu Wachstum und Ernährung*. R. Oldenburg, München-Berlin

Rubner M (1909). *Kraft und Stoff in Haushalte der Natur*. Akademische Verlagsgesellschaft, Leipzig

Sacher GA (1959). Relation of life span to brain weight and body weight. In: Wolstenholme GEW, O'Connor M (eds) *The Life Span of Animals*. Little, Brown, Boston, pp 115–141

Schmidt-Nielsen K (1984). *Scaling: Why Is Animal Size So Important?* Cambridge University Press, Cambridge

Scholander PF, Hock R, Walters V, Johnson F, Irving L (1950). Heat regulation in some arctic and tropical animals and birds. *Biol Bull* 99: 237–258

Index

DATE DUE

MAY 1 8 1998			
JUN 0 2 1998			

DEMCO 38-297